THE ORIGIN OF
THE CHEMICAL ELEMENTS

THE WYKEHAM SCIENCE SERIES

General Editors:

PROFESSOR SIR NEVILL MOTT, F.R.S.
Emeritus Cavendish Professor of Physics
University of Cambridge

G. R. NOAKES
Formerly Senior Physics Master
Uppingham School

To introduce the present state of science as a university subject to students approaching or starting their university careers is the aim of the Wykeham Science Series.   Each book seeks to reinforce the link between school and university levels, and the main author, a university teacher or research worker distinguished in the field, is assisted by an experienced sixth-form schoolmaster.

THE ORIGIN OF
THE CHEMICAL ELEMENTS

R. J. Tayler

University of Sussex

DISTRIBUTED BY
SPRINGER-VERLAG NEW YORK INC.
175 Fifth Avenue New York, NY 10010
ISBN #91100-6

WYKEHAM PUBLICATIONS (LONDON) LTD
(A MEMBER OF THE TAYLOR & FRANCIS GROUP)
LONDON AND WINCHESTER
1972

First published 1972 by Wykeham Publications (London) Ltd.

Cover illustration—The Crab Nebula, which is the remnant of the supernova of AD 1054 (Photograph, Hale Observatories).

ISBN 0 85109 280 2

Printed in Great Britain by Taylor & Francis Ltd.
10–14 Macklin Street, London, WC2B 5NF

Distribution:

UNITED KINGDOM, EUROPE, MIDDLE EAST AND AFRICA
Chapman & Hall Ltd. (a member of Associated Book Publishers Ltd.), 11 New Fetter Lane, London, EC4P 4EE and North Way, Andover, Hampshire.

WESTERN HEMISPHERE
Springer-Verlag New York Inc., 175 Fifth Avenue, New York, New York 10010

AUSTRALIA AND NEW GUINEA
Hicks Smith & Sons Pty. Ltd., 301 Kent Street, Sydney, N.S.W. 2000.

NEW ZEALAND AND FIJI
Hicks Smith & Sons Ltd., 238 Wakefield Street, Wellington.

ALL OTHER TERRITORIES
Taylor & Francis Ltd., 10–14 Macklin Street, London, WC2B 5NF.

PREFACE

DURING this century study of radioactivity and the discovery of nuclear fusion and fission reactions has demonstrated that the chemical composition of the Universe changes with time. As most of the material in the Universe appears to be either hydrogen or helium and as fusion reactions provide the energy which stars radiate and gradually convert light elements into heavier elements, there is a possibility that the Universe originally contained only light elements and that all of the heavy elements have been produced by nuclear reactions in stars.

This book describes what is at present known about the evolution of the chemical composition of the Universe. This is a very complicated subject, which involves many branches of astronomy and physics, and which is at an active stage in its development. I have, therefore, tried to describe in broad outline the considerable progress which has been made, without disguising the many uncertainties which remain.

It is impossible in a short book to give credit for all of the individual advances in the subject and there are few names in the text. I am grateful to many colleagues for permission to use their results, particularly in my diagrams. Amongst the astronomers and physicists who have made outstanding contributions to the subject in the last twenty years are E. M. Burbidge, G. R. Burbidge, A. G. W. Cameron, W. A. Fowler, F. Hoyle and E. E. Salpeter on the theoretical side, and J. L. Greenstein, H. E. Suess, A. Unsöld and H. C. Urey on the observational side. My own interest in the subject owes much to William Fowler, Fred Hoyle and Bernard Pagel.

Apart from the use of some special units, such as parsec and electron volt, all numerical quantities are expressed in SI Units. These are explained, for example, in the Royal Society booklet *Symbols*, *Signs and Abbreviations* (1969) and the abbreviation of units and the labelling of graph axes should be especially noted. A list of the more important symbols used in the book and of numerical values of physical constants to the accuracy needed is given on pages vii and viii.

I am grateful to Mr. Douglas Meyer for drawing all of the figures and to Mrs. Hazel Freeman for her careful typing of the manuscript. I am once again especially indebted to my schoolmaster collaborator, Mr. Alan Everest, for his many suggestions and criticisms which have led to considerable improvements in the book.

Lewes, *March* 1972   R. J. Tayler

SYMBOLS

A	number of nucleons in nucleus
B	magnetic induction
$B_\nu(T)$	Planck function
$E, \mathscr{E}$	energy
f	f-value of atomic transition
H	Hubble's constant
I, I_ν	intensity of radiation
L	luminosity
m	particle mass
M	mass
n	number of particles in cubic metre
n, N	abundance of nuclear species
N	number of neutrons in nucleus
P	pressure, production rate of isotope
Q	binding energy of nucleus
r, R	distance
t, T	time
T	temperature
T_e	effective temperature
U, B, V	photoelectric stellar magnitudes
v, V	velocity
W	equivalent width of spectral line
Z	number of protons in nucleus
λ	wavelength, decay rate of nucleus
μ	mean molecular weight
ν	frequency
ρ	density
σ	cross-section
τ	neutron exposure
e^-, e^+	electron, positron
d	deuteron
n	neutron
p	proton
γ	photon
$\nu, \bar{\nu}$	neutrino, anti-neutrino

Suffixes s and $\odot$ refer to values at the surface of a star and solar values respectively. The symbol for an isotope, such as $^{238}_{92}U$, is sometimes written ^{238}U or $_{92}U$ if the emphasis is on the mass number or the charge number respectively.

NUMERICAL VALUES

Fundamental Physical Constants

a	radiation density constant	$7 \cdot 55 \times 10^{-16}$ J m^{3}K^{-4}
c	velocity of light	$3 \cdot 00 \times 10^8$ m s^{-1}
e	charge on electron	$1 \cdot 60 \times 10^{-19}$C
G	gravitational constant	$6 \cdot 67 \times 10^{-11}$N m^2 kg^{-2}
h	Planck's constant	$6 \cdot 62 \times 10^{-34}$J s
k	Boltzmann's constant	$1 \cdot 38 \times 10^{-23}$J K^{-1}
m_e	mass of electron	$9 \cdot 11 \times 10^{-31}$kg
m_H	mass of hydrogen atom	$1 \cdot 67 \times 10^{-27}$kg
$\mathscr{R}$	gas constant (k/m_H)	$8 \cdot 26 \times 10^3$J K^{-1}kg^{-1}

Astronomical Quantities

$L_\odot$	luminosity of Sun	$3 \cdot 90 \times 10^{26}$ W
$M_\odot$	mass of Sun	$1 \cdot 99 \times 10^{30}$ kg
$r_\odot$	radius of Sun	$6 \cdot 96 \times 10^8$ m
$T_{e\odot}$	effective temperature of Sun	5780 K

Non SI Units

barn (unit of cross-section)	10^{-28} m^2
electron volt (unit of energy)	$1 \cdot 60 \times 10^{-19}$J
light year (unit of distance)	$9 \cdot 5 \times 10^{15}$ m
parsec (unit of distance)	$3 \cdot 09 \times 10^{16}$ m
year	$3 \cdot 16 \times 10^7$ s
Ångstrom, Å (unit of length)	

CONTENTS

Preface   v

Symbols   vii

Numerical Values   viii

Chapter 1   INTRODUCTION AND ASTRONOMICAL
              BACKGROUND   1

Chapter 2   ELEMENT ABUNDANCES IN THE
              SOLAR SYSTEM   18

Chapter 3   CHEMICAL COMPOSITION OF THE SUN,
              STARS AND GASEOUS NEBULAE   29

Chapter 4   ELEMENT ABUNDANCES IN COSMIC
              RAYS   55

Chapter 5   THEORETICAL BACKGROUND TO THE
              ORIGIN OF THE ELEMENTS   65

Chapter 6   COSMOLOGICAL ELEMENT PRODUCTION   93

Chapter 7   ELEMENT PRODUCTION IN THE
              GALAXY   108

Chapter 8   RADIOACTIVE CHRONOLOGIES   146

Chapter 9   CONCLUDING REMARKS   157

Appendix   THERMODYNAMIC EQUILIBRIUM   162

Index   165

Suggestions for further reading   168

The Wykeham Series   170

CHAPTER 1
introduction and astronomical background

Elements, atoms, nuclei and elementary particles

For the whole of the recorded history of science, there has been concern about the nature of the basic building bricks of the physical Universe and with the related problem of creation. This book is, therefore, about the impact of modern developments in astronomy and physics on one of the oldest problems in science and philosophy. The history of the subject has included some periods of great simplicity, when there were thought to be very few basic types of matter, but at other times the subject has seemed to be much more complex.

Aristotle held the view that there were four basic *elements*, fire, air, water and earth and he believed that all the substances with which we are familiar on Earth were different combinations of these four elements alone. Celestial objects were supposed to be composed of a fifth more perfect substance. Belief in the four elements continued until late into the Middle Ages. Because it was believed that all metals were combinations of them, alchemists hoped that it would be possible to change the combinations and to transmute base metals into gold.

Along with discussions about the nature of the ultimate elements, there was dispute about whether matter was infinitely divisible or whether eventually for any element a particle would be reached which could not be divided any further; the latter is the atomistic view first propounded by Democritus and Lucretius. Eventually in the nineteenth century it appeared that the atomistic view had prevailed and that the ultimate building bricks were atoms of what we now call the chemical elements; as there were almost 100 of these, the basic building bricks were thought to be very numerous.

The periodic table

The idea that the atoms of the chemical elements were indivisible did not survive very long. Its ultimate replacement was probably foreshadowed by the development of the *periodic table* of the elements by Newlands, Meyer and Mendeleev (table 1). Mendeleev showed that groups of elements had similar chemical properties and that, if the elements were arranged in order of atomic weight, elements with similar properties occurred at periodic intervals. Such regularity

1

1a	2a	3b	4b	5b	6b	7b	8	8	8	1b	2b	3a	4a	5a	6a	7a	0
$_{1}$H																	$_{2}$He
$_{3}$Li	$_{4}$Be											$_{5}$B	$_{6}$C	$_{7}$N	$_{8}$O	$_{9}$F	$_{10}$Ne
$_{11}$Na	$_{12}$Mg											$_{13}$Al	$_{14}$Si	$_{15}$P	$_{16}$S	$_{17}$Cl	$_{18}$Ar
$_{19}$K	$_{20}$Ca	$_{21}$Sc	$_{22}$Ti	$_{23}$V	$_{24}$Cr	$_{25}$Mn	$_{26}$Fe	$_{27}$Co	$_{28}$Ni	$_{29}$Cu	$_{30}$Zn	$_{31}$Ga	$_{32}$Ge	$_{33}$As	$_{34}$Se	$_{35}$Br	$_{36}$Kr
$_{37}$Rb	$_{38}$Sr	$_{39}$Y	$_{40}$Zr	$_{41}$Nb	$_{42}$Mo	$_{43}$Tc*	$_{44}$Ru	$_{45}$Rh	$_{46}$Pd	$_{47}$Ag	$_{48}$Cd	$_{49}$In	$_{50}$Sn	$_{51}$Sb	$_{52}$Te	$_{53}$I	$_{54}$Xe
$_{55}$Cs	$_{56}$Ba	$_{57}$La†	$_{72}$Hf	$_{73}$Ta	$_{74}$W	$_{75}$Re	$_{76}$Os	$_{77}$Ir	$_{78}$Pt	$_{79}$Au	$_{80}$Hg	$_{81}$Tl	$_{82}$Pb	$_{83}$Bi	$_{84}$Po*	$_{85}$At*	$_{86}$Rn*
$_{87}$Fr*	$_{88}$Ra*	$_{89}$Ac*‡															

†Lanthanides	$_{58}$Ce	$_{59}$Pr	$_{60}$Nd	$_{61}$Pm	$_{62}$Sm	$_{63}$Eu	$_{64}$Gd	$_{65}$Tb	$_{66}$Dy	$_{67}$Ho	$_{68}$Er	$_{69}$Tm	$_{70}$Yb	$_{71}$Lu
‡Actinides	$_{90}$Th*	$_{91}$Pa*	$_{92}$U*	$_{93}$Np*	$_{94}$Pu*	$_{95}$Am*	$_{96}$Cm*	$_{97}$Bk*	$_{98}$Cf*	$_{99}$Es*	$_{100}$Fm*	$_{101}$Md*	$_{102}$No*	$_{103}$Lw*

Table 1. The periodic table of the chemical elements. Elements appearing in the same column possess similarities in their chemical properties. The atomic number is shown with the atomic symbol. Those elements marked with an asterisk possess no stable isotopes.

would be surprising if each atom were a totally independent building brick, but not if the atoms possessed some substructure with regular properties. This substructure was revealed when J. J. Thomson discovered the electron, when Lord Rutherford showed that an atom possessed a very small nucleus surrounded by a cloud of electrons occupying a much larger volume and when Niels Bohr's theory of the nuclear atom showed that in each type of atom the electrons occupied a characteristic set of energy states. The spectral lines of elements were then produced when an electron changed its energy state and released energy in the form of radiation.

Radioactivity, isotopes and nuclear structure

The discovery of natural radioactivity, changing uranium and thorium into lead, for example, showed that atoms not only possessed substructure, but that they could not be regarded as permanent. It was soon realised that radioactivity involved modification of an atomic nucleus. Eventually after the discovery of the neutron in 1932, it appeared that the ultimate building bricks of ordinary matter were negatively charged electrons, positively charged protons about 1840 times more massive than electrons and uncharged neutrons with almost the same mass as protons; in addition it appeared that in some sense a neutron could be formed out of a proton and an electron. A nucleus was made up of protons and neutrons bound together by a nuclear force whose properties were then largely unknown. Some chemical elements had previously been shown to exist in several forms which had identical chemical properties but different atomic masses. The existence of such *isotopes* could now be explained. A nucleus of a heavy isotope contained the same number of protons but more neutrons than a light isotope; see fig. 1. Chemical properties were determined by the atomic electrons of which, in the neutral atom, there were as many as there were protons in the nucleus.

Fig. 1. The stable isotopes of helium. ³He contains only one
neutron while ⁴He contains two neutrons.

Anti-matter and elementary particles

Already when this picture of matter composed of protons, neutrons and electrons was being produced, it was clear that they were *not* the only fundamental particles. The discovery of the positron with the same mass as the electron but with a positive charge foreshadowed the eventual detection of the antiproton and antineutron*; they are examples of what is known now as *antimatter*. Positrons, antiprotons and antineutrons are not present in ordinary matter but it is not obvious why our part of the Universe could not be made of antimatter rather than matter. Indeed it is possible that some regions in the Universe *are* composed of antimatter. It would not be possible to detect them from the light they emit as the photon is an exceptional particle which appears to be identical with its antiparticle. As particles and antiparticles annihilate one another to produce photons, intense radiation might be received from regions where matter and antimatter are in contact. If there is a substantial amount of anti-matter in the Universe, it is important to ask how the matter and antimatter became separated; if there are no such regions, it is interesting to ask why the Universe is made of matter rather than antimatter.

In the past 40 years, many more *fundamental particles* have been discovered and they now number many hundreds. They include the neutrino, which is a particle without mass or charge†, the μ-meson, which appears to be identical with the electron except that it is about 200 times as heavy, the π-meson, which plays an important role in the structure of the nucleus, and many other particles that are heavier versions of the proton and neutron. The subject of elementary particle physics is in a very confused state at the moment, although there are gropings towards arrangements of the elementary particles, which are reminiscent of the periodic table of the chemical elements. This has, in turn, led to a suggestion that protons and neutrons are themselves composed of three as yet undiscovered particles which have been given the name of *quarks*.

Although it is not at present clear what the true fundamental particles of nature are, in many respects protons, neutrons and electrons can still be regarded as the fundamental particles. Most of the recently discovered elementary particles, which have been produced in particle accelerators, are very unstable and decay into other particles, many decays having a half life as short as 10^{-23} s‡. Even the neutron is unstable if it is not bound to another particle, and decays into a proton, electron and antineutrino with a half life of

* It is now known that every elementary particle possesses an antiparticle which has equal mass but equal and opposite values for other basic properties such as electric charge.

† It shares these properties with the photon and they both move with the speed of light but they differ in other respects; in particular it has an intrinsic spin, similar to the electron, but it always spins in the same sense about its direction of motion.

‡ The negatively charged π-meson, π^-, for example, can decay into an electron and an antineutrino with a mean life of $2 \cdot 5 \times 10^{-8}$ s. The uncharged π-meson decays into photons with a mean life of $2 \cdot 2 \times 10^{-16}$ s.

about ten minutes, but when it is bound into a nucleus it gains stability. It appears then that most matter is made up of protons, neutrons and electrons. In the remainder of this book we do not discuss problems of elementary particle physics and antimatter any further but we ask *why these protons, neutrons and electrons are arranged in the particular chemical elements in the quantities which we observe and whether there has been any significant change in the chemical composition of the Universe during its lifetime.*

Nuclear reactions in stars

We have mentioned above that the existence of natural radioactivity shows that some change in chemical composition is occurring but in our everyday experience we do regard the atoms of the non-radioactive elements as immutable. Materials undergo physical changes such as melting and vaporization and the arrangement of atoms in molecules can be altered, but the properties of the *individual atoms* are unchanged. However, it is known that there are some important changes of chemical composition in the Universe. It has been realised for some time that the only source of energy which can have enabled the Sun to radiate steadily throughout the Earth's geological history is nuclear fusion reactions converting hydrogen into helium. This means that there is a gradual change in the chemical composition of the Sun. For reasons which have been described in the companion book in this series, *The Stars: their structure and evolution**, and which will be discussed further below, it is believed that there is a gradual conversion of light elements into heavy elements as a star evolves, the conversion of hydrogen into helium being followed by the production of carbon, oxygen, neon, magnesium, silicon and other heavier elements. As observations of stars and gas clouds suggest that most of the material in that part of the Universe which we can study is hydrogen or helium, there exists the possibility that the originally created matter had a very simple chemical composition and that all of the complications have been produced since then.

The central question with which this book is concerned is thus: *what can observations of the present composition of the stars and gas clouds in our neighbourhood in the Universe, combined with theories of the transmutation of elements in these objects, tell us about the original chemical composition of the Universe?*

At present it is not possible to give a clear answer to this question. To discuss it adequately many interesting problems in astronomy must be studied. Considerable progress has, however, been made and the main aim of this book will be to explain all of the basic

---

* There will be frequent references to this book in what follows. It will usually simply be called *The Stars*. Although it would be an advantage to have read *The Stars* before the present book, all results from it which are needed will be explained briefly.

concepts involved and to outline this progress. We cannot expect to deduce the initial composition of the Universe by working back from the present complicated and uncertain observations. Instead, it is convenient to work in the opposite direction. If we assume an initial composition, we can try to calculate what the present composition should be, and we hope to discover a (simple) initial composition which leads to something like the composition observed today.

Related branches of astronomy

The subject of this book is closely related to three other branches of astronomy. These are:
   (i) cosmology;
  (ii) the structure and evolution of stars (and any other objects in which heavy elements are produced);
 (iii) the structure and evolution of galaxies.

Cosmology is concerned with the structure of the Universe on the largest possible scale and with questions of origin and creation. Any cosmological theory which attempts to discuss the entire life history of the Universe must make assumptions about the initial distribution of the chemical elements. In Chapter 6 we shall discuss two particularly simple cosmological theories, the *big-bang theory* and the *steady state theory*. We will describe what they have to say about the chemical composition of the Universe and will consider reasons unconnected with the chemical composition which may cause the theories to be untenable.

As we have already mentioned, nuclear transmutation occurs in stars and this means that we must be interested in the structure and evolution of stars of all masses. This is the subject matter of *The Stars* and we shall frequently quote results from it in what follows. Later we shall also see that it has recently been suggested that objects much more massive than ordinary stars may have played an important role in nucleosynthesis and the structure of such objects must also be studied. In discussing the production of heavy elements from light elements by nuclear fusion reactions inside stars, we immediately encounter one of the basic difficulties of the subject. Nuclear reactions occur most readily at high temperatures and, in stars, the highest temperatures occur near their centres. In contrast, the light which we receive from stars gives information about the chemical composition of the *surface* layers of the stars only. This means that the heavy elements produced in stars can only be observed if there exist processes which carry the heavy elements from the centre to the surface or into interstellar space. It proves much more difficult to study the problem of *mass loss* from stars than the evolution of stars at constant mass.

In fact, we believe that the heavy elements which we observe in stars today have been produced in other stars earlier in the galactic

6

lifetime, that they have been expelled into interstellar space from those stars and have then been incorporated into a new generation of stars. A discussion of the exchange of mass between stars and the interstellar medium, in the form of mass loss from stars and the formation of new stars, and of the gradual change with time of the chemical composition of the interstellar gas and hence of the stars formed from it, implies an interest in the third subject listed above; the evolution of galaxies and particularly of our Galaxy.

Although most branches of physics enter into our subject through study of the structure of stars and galaxies, it is nuclear physics which is most directly concerned with the processes converting one element into another. A purely theoretical study of the origin of the chemical elements involves three major steps. In the first place we must enumerate the nuclear processes involved and discuss the physical conditions required for each to be important. Second we must decide what objects in the Universe possess the conditions relevant to each process. Finally we must discuss the observed number of objects of each type and must attempt to estimate the overall rate of production of each chemical element. At the present time each step in this discussion is only imperfectly understood, the greatest uncertainties being in the final step of the argument.

Constituents of the Universe

We now turn to a brief description of the major constituents in the Universe. Before we do this it is useful to introduce some units of distance which are more useful than the metre in discussing astronomical distances. One convenient unit is the distance travelled by light in one year.

$$1 \text{ light year} = 9 \cdot 5 \times 10^{15} \text{ m}. \tag{1.1}$$

Professional astronomers normally use a distance known as the parsec, where

$$1 \text{ parsec} = 3 \cdot 09 \times 10^{16} \text{ m} = 3 \cdot 26 \text{ light years}. \tag{1.2}$$

Both of these units will be used in this book; though it is perhaps easier to obtain a rapid understanding of distances expressed in light years. As a unit of mass throughout the book we take the solar mass $(M_\odot)$, which is

$$M_\odot = 1 \cdot 99 \times 10^{30} \text{ kg}. \tag{1.3}$$

To a first approximation we can say that the main constituents in the Universe are galaxies, which themselves have as principal constituents stars and gas clouds. The galaxies vary considerably in size but a large galaxy probably contains between 10^{11} and 10^{12} stars and has a mass of more than $10^{11} M_\odot$. Galaxies are of the order of 10^6 light years apart although they are not uniformly distributed through the Universe and definite clusters of galaxies exist. The most distant galaxies observed in large telescopes are several times

10^9 light years distant from us; this means the number of galaxies in the observable Universe is certainly measured in thousands of millions.

The structure of the Galaxy

Although we have several times referred to the chemical composition of the Universe, most of the observations refer to the galaxy in which the Sun is situated which is known as *the Galaxy*. The structure of the Galaxy is shown schematically in figs. 2 and 3. From the side

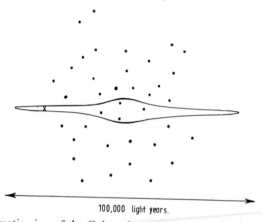

100,000 light years.

Fig. 2.   Schematic view of the Galaxy from the side, showing the thin galactic disk and the central nuclear bulge.   The position of the Sun is marked with a cross and the filled circles represent globular clusters.

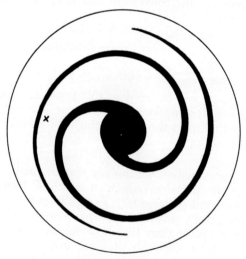

Fig. 3.   A schematic view of the Galaxy from above.   The spiral structure is shown and the position of the Sun is marked with a cross.

the Galaxy has a very flattened appearance and most of the mass of the Galaxy is contained in the flat *disk* and central *nucleus*, but there is a smaller amount in a spherical *halo*. The galactic diameter is of order 10^5 light years and the thickness of the disk is little more than 10^3 light years. The total mass is believed to be between 10^{11} and 2×10^{11} $M_\odot$ and we shall usually use the value 1.5×10^{11} $M_\odot$. The major constituent of the Galaxy is stars, of which there are believed to be a few times 10^{11}, but 5 or possibly 10 per cent of the galactic mass is

Fig. 4. The spiral galaxy NGC 4565 seen edge on.
(*Photograph from the Hale Observatories.*)

9

interstellar gas and dust. The gas and dust is concentrated in the disk of the Galaxy and particularly in the spiral arms, which are shown in fig. 3. As we are inside the Galaxy, our knowledge of what it would look like from outside is obtained indirectly but in fig. 4 is shown a photograph of a galaxy which is believed to be similar.

Many of the stars in the Galaxy are members of star clusters. These are physical groupings of stars, whose mutual gravitational attraction is stronger than disruptive forces due to the remainder of the Galaxy. These are of two types, globular clusters and galactic clusters. Globular clusters, which have a compact circular appearance, are spread throughout the Galaxy including the halo and they contain between 10^5 and 10^6 stars each. Galactic clusters, which are less compact and contain many fewer stars, are found in the disk. The most reliable information about chemical compositions refers to the solar neighbourhood in the Galaxy which is a region of radius the order of 10^3 light years centred on the Sun; the position of the Sun is marked in figs. 2 and 3.

Properties of stars

The observed properties of stars have been discussed in considerable detail in *The Stars*. Here we merely list those properties which are of particular importance in this book. The principal observed properties of stars are *luminosity*, L_s, and *surface temperature*, T_s. The luminosity is the total energy emitted by the star per second and this can be deduced from the energy received on Earth if the distance of the star is known. It is possible to deduce the surface temperature of a star from the frequency distribution of the radiation which it emits. If the luminosities of stars of known distance are plotted against their surface temperatures, fig. 5 results. This is known as the *Hertzsprung-Russell* diagram. Stars lie in fairly well defined regions of the diagram; more than 90 per cent of all stars are probably *main sequence stars*, the *white dwarfs* are the second most numerous class and both *giants* and *supergiants* are much rarer. It is explained in *The Stars* that a star can be a main sequence star at one stage in its life history and a giant or a white dwarf at another stage. Thus, the majority of all stars are main sequence stars because the main sequence phase is the longest stage in stellar evolution; the one in which nuclear reactions converting hydrogen into helium are supplying the energy radiated by the star.

The mass, M_s, of some stars which are partners in double star systems can be determined and the masses and luminosities of main sequence stars have the property that the luminosity increases as a rather high power of the mass. This *mass-luminosity relation* is shown in fig. 6. The radius, r_s, can only be measured for a very small number of stars. Theoretical astronomers normally describe stars in terms of the *effective temperature*, T_e, instead of the surface

10

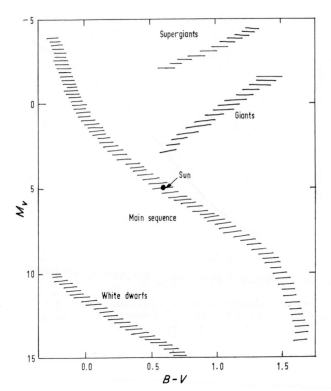

Fig. 5. The Hertzsprung-Russell diagram for nearby stars. The visual magnitude M_V is plotted against colour index $B-V$ and most stars fall in four well-defined groups. (M_V is proportional to $-\log L_s$ and $B-V$ is related to $-\log T_s$; the hottest brightest stars are in the top left of the diagram.)

temperature. T_e is defined as the temperature of a *black body*, which has the same radius and luminosity as the star, and it satisfies

$$L_s = \pi a c r_s^2 T_e^4, \tag{1.4}$$

where a is the radiation density constant ($7 \cdot 55 \times 10^{-16} \mathrm{Jm^3K^{-4}}$) and c is the velocity of light ($3 \cdot 00 \times 10^8 \mathrm{m\ s^{-1}}$). T_e can be calculated more easily than T_s and, although stars do not radiate like black bodies, T_s and T_e usually differ by only a few per cent.

The properties of the Sun, including its mass which has already been stated in eqn. (1.3), are

$$\left. \begin{aligned} M_\odot &= 1 \cdot 99 \times 10^{30} \mathrm{\ kg}, \\ L_\odot &= 3 \cdot 90 \times 10^{26} \mathrm{\ W}, \\ r_\odot &= 6 \cdot 96 \times 10^8 \mathrm{\ m}, \\ T_{e\odot} &= 5780 \mathrm{\ K}. \end{aligned} \right\} \tag{1.5}$$

11

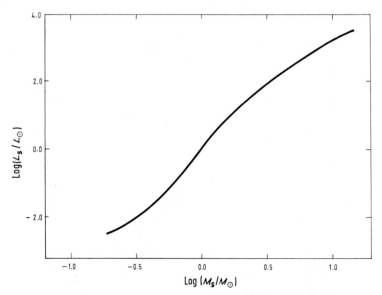

Fig. 6. The mass-luminosity relation. The luminosity L_s is plotted against mass M_s. $L_\odot$ and $M_\odot$ are the luminosity and mass of the Sun. Stars with accurately known luminosity and mass lie close to the curve shown.

We can now state, in terms of the solar values, what ranges of values for M_s, L_s, r_s and T_e have been found in other stars. These are approximately

$$\left.\begin{array}{l} 10^{-1}\, M_\odot \;< M_s < 50\, M_\odot, \\ 10^{-4}\, L_\odot \;< L_s \;< 10^6\, L_\odot, \\ 10^{-2}\, r_\odot \;\;\;< r_s \;\;< 10^3\, r_\odot, \\ 2 \times 10^3\, \text{K} < T_e \;< 10^5\, \text{K}. \end{array}\right\} \qquad (1.6)$$

The very high luminosities of exploding supernovae have been excluded from these limits and it is very likely indeed that stars exist with masses, radii and luminosities smaller than those shown in inequalities (1.6).

The white dwarfs have a very low luminosity and from eqn. (1.4) their radii must be small. A few white dwarf masses are known and these are comparable with the mass of the Sun. This implies that this type of star has a very high mean density of order 10^{10} kg m^{-3}. They are believed to be stars which are in the final stages of stellar evolution. Theoretical astronomers have predicted the existence of another type of dying star, the *neutron star*, which is essentially made of neutrons and has a central density of order 10^{18} kg m^{-3}. Neutron stars are not shown in fig. 5 as they have not definitely been observed but on page 122 we shall see why they are believed to exist. White dwarfs and neutron stars should eventually cool down to a very low

12

temperature but theory predicts there is a maximum mass for both white dwarfs and neutron stars; these are apparently $\sim 1\cdot 2\ M_\odot$ and $\sim 2\ M_\odot$ respectively. Above the critical mass the attractive force of gravity forces the material of the star to an even higher density as its temperature falls towards absolute zero and, if the presently accepted laws of physics are true, a star of greater mass than $2\ M_\odot$ cannot become a dead star of finite radius but collapses to a singularity and becomes what is known as a *black hole*. The existence of black holes is postulated but not certain and these will be discussed further in Chapter 7.

Observations of element abundances
To a large extent, work in the subject of this book has concentrated on trying to obtain a history of the chemical composition of the Galaxy. How is progress made within this framework? The first necessity is the demonstration that the chemical composition of all stars and gas clouds in the Galaxy is not the same. There *is* in fact a considerable degree of uniformity. Most stars share the property that they are largely composed of hydrogen and helium and that the relative abundances of the heavier elements have the general character shown in fig. 7. There are, however, some important differences from star to star in the total amount of heavy elements in proportion to light elements and some differences in the relative abundances of various heavy elements. As the general character of fig. 7 can be seen in the abundances in most stars, it is to be hoped that this will

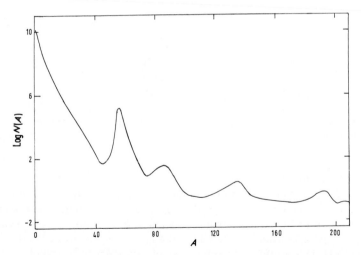

Fig. 7. A schematic abundance curve. A is the atomic mass number and $N(A)$ is the number of atoms with mass number A; the actual numbers are chosen so that there are 10^6 silicon atoms. The true abundance curve is much more irregular.

13

give information about the *basic processes* involved in the origin of the elements. In contrast, the variations of chemical composition should tell us something about the *evolution* of chemical composition of the Galaxy.

Thus, having shown that differences of chemical composition exist, it is necessary to see whether there is any pattern in the differences. Two obvious possibilities arise. If the chemical composition of the Galaxy varied from place to place when it was formed or if the production of heavy elements has been concentrated in particular regions of the Galaxy, we should expect to find a correlation between stellar chemical composition and the place of origin of the star. Note that, because all stars and gas clouds in the Galaxy move under the gravitational attraction of all other objects in the Galaxy, the place of origin may differ from the present position. Secondly, if there has been a fairly steady production of heavy elements throughout the Galaxy in the galactic lifetime, we can expect chemical composition to be correlated with the age of objects. As has been described in *The Stars*, it is possible to estimate the ages of star clusters in the Galaxy and it therefore makes sense to study their chemical composition as a function of both *age* and *position*.

The first correlation which became apparent from study of stars in the solar neighbourhood in the Galaxy was that the heavy element content of the interstellar medium must have increased with time as the recently formed stars contain more heavy elements than older stars. As only the *very oldest* stars contain a substantially lower amount of heavy elements than those recently formed, it seems that the rate of production of heavy elements has not been uniform in time; there must have been a much more rapid rate of production of heavy elements in the early life of the Galaxy than there is at present. For some time it appeared that this was a simple relationship, but more recently it has become clear that chemical composition may also vary significantly from place to place within the Galaxy; in particular, the concentration of heavy elements may be higher near the centre of the Galaxy than in its outer regions. These points are discussed further in Chapter 3.

Sites of nucleosynthesis

It has already been mentioned that it is usually supposed that the heavy elements have been produced in stars and that the material produced must be expelled from the stars so that it can be incorporated in a new generation of stars. The one type of event in the Galaxy in which a large amount of material has been shown to be expelled by a star is the explosion of a supernova. It seems likely that several supernovae explode in a typical large galaxy in each century and that a considerable amount of heavy elements could be produced in a star,

14

either before it became a supernova or during the explosion, and be expelled into space. Supernovae *may* be the main source of the heavy elements and this question is discussed in considerable detail in Chapter 7.

Although the most violent event that has been observed in our Galaxy is the explosion of a supernova, it has become apparent in recent years that much more violent explosions have occurred in the central regions of some galaxies. These explosions include those which are believed to produce the double radio sources. The development of radio astronomy has led to the discovery of many objects in the Universe which are strong emitters of radio waves. Included amongst these *radio sources* are radio galaxies, many of which are double sources of the type illustrated in fig. 8. With a low resolution radio telescope it appears that a galaxy which is already known optically is in addition a strong source of radio waves. When the galaxy is studied with higher resolution it is found that the radio emission is concentrated in two regions on either side of the optical galaxy and often some considerable distance away from it.

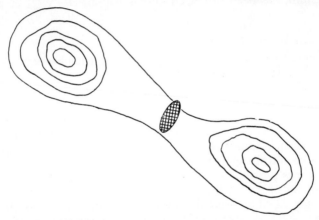

Fig. 8. A double radio source. The cross-hatched area is an elliptical galaxy. On either side of it are regions of strong radio emission. Schematic contours of equal radio emission are shown; the strongest emission comes from the centres of the two components.

There is at present no completely accepted theory of the origin of these *double* radio sources, but it is generally agreed that there has been an explosion in the centre of the galaxy and that two masses have been hurled out of the galaxy in opposite directions. In order to produce the observed radio emission, it is almost certainly necessary that the masses thrown out contain very energetic charged particles and a magnetic field and that the radio waves are produced by the charged particles which radiate when they move in a magnetic field.

It is possible to make an estimate of the total amount of energy which must be contained in the radiating regions and to deduce that the original explosion must have been very violent indeed and that it must have involved a mass of $10^8 M_\odot$ or even more.

As these explosions are so much more violent than supernovae and as it appears that very high energy particles are produced in them, it seems quite possible that nucleosynthesis also occurs in the explosion. It is possible that such explosions occur in most galaxies and the indication that the heavy element content in our Galaxy may be higher near to the centre than elsewhere has led to suggestions that this has been caused not by a concentration of supernovae near the galactic centre but by a small number of explosions of more massive objects. This has led in turn to an interest in the structure and evolution of objects much more massive than stars and to a discussion of the nucleosynthesis that occurs in them. At the moment the relative importance of stars and more massive objects in the origin of the chemical elements is rather unclear, although present calculations indicate that it is more likely that helium, rather than elements heavier than helium, has been produced in massive objects.

Outline of contents of book

In the next three chapters we shall consider the observations of the relative abundances of the different chemical elements in the Universe. Such observations are of three types. Samples of material from the Earth, Moon and meteorites can be studied directly and can be analysed by chemical methods. Only limited regions of the Earth, Moon and meteorites are accessible to direct observation, which means that an estimate must be made of the possible segregation of different chemical elements in different parts of the Earth and Moon. Even so, it is possible to make a reasonable estimate of the chemical compositions of the Earth, Moon and meteorites and this is discussed in Chapter 2. The second type of observation involves the interpretation of the spectrum of electromagnetic radiation originating in stars and gas clouds. The presence of absorption and emission lines in these spectra indicates that particular chemical elements are present in the objects. It is thus relatively easy to see that an element is present, but it can be much more difficult to obtain a good estimate of the quantity of the element that is present. The *deduction* of abundances from spectra is discussed in Chapter 3. Finally it is possible to discover something about the chemical composition of the nuclei in cosmic rays, the very high energy charged particles which reach the Earth from all directions in space, and this is discussed in Chapter 4.

After discussing the observed abundances, we turn to a consideration of the processes by which these abundances have been produced.

In Chapter 5, we describe the processes without considering in any detail the astronomical objects in which they might occur. It is possible to see that most elements and isotopes can have been produced either by nuclear fusion reactions or by processes involving the capture of neutrons by existing relatively heavy elements. There are two small groups of isotopes which are not produced by any of the main processes but other suggestions are made for their production mechanisms.

In Chapters 6 and 7 we discuss where the nuclear reactions described in Chapter 5 may have occurred during the lifetime of the Universe and what is the relation of the present composition to the original composition. It appears plausible that the initial composition was either pure hydrogen or a mixture of hydrogen and helium and observations which may enable us to decide between these possibilities are discussed in Chapter 6. Nuclear reactions in stars and more massive objects and their possible influence on the chemical composition of the Galaxy are studied in Chapter 7. The radioactive elements in the Earth, Moon and meteorites can be used to obtain information about both the age of the solar system and the past rate of formation of the heavy elements and these *radioactive chronologies* are described in Chapter 8.

It cannot be stressed too clearly that the Origin of the Chemical Elements is an active research subject and not one that is fully understood. For this reason most of the conclusions described in this book are only tentative and they will probably be modified by future observational and theoretical work. In addition the subject is far too large to be discussed comprehensively in a short book of 164 pages. Nevertheless, it should be apparent why it is regarded as a very exciting subject by those who work on it. The final chapter of the book is devoted to a discussion of some of the problems which remain to be solved in future.

CHAPTER 2
element abundances in the solar system

Introduction

IN the present chapter we shall discuss the chemical composition of objects in the solar system—excluding the Sun, which is studied in the same way as other stars and which will be discussed in the next chapter.

In the case of the Earth, meteorites, and now to a limited extent, the Moon, it is possible to determine the composition directly by *chemical analysis*. This statement should perhaps be qualified immediately as we do not have direct access to the interior of the Earth and meteorites lose quite a large amount of their material in their passage through the Earth's atmosphere. Direct chemical analysis has two great advantages: (i) element abundances can be obtained which are free from the uncertainties involved in the interpretation of spectral lines, which will be discussed in the next chapter, and (ii) *isotopic ratios* can be obtained. The spectral lines produced by different isotopes of the same element differ only very slightly in frequency, except in the case of very light elements. Because actual spectral lines are broadened over a range of frequencies, as will be described in Chapter 3, in most cases it is not possible to determine which *isotopes* of a particular element are present in stars.

As we shall see in Chapter 5, theories of nucleosynthesis predict *isotopic* as well as *elemental* abundances and it is therefore useful to have some observed isotopic abundances to compare with theory. In Chapter 5 we shall see that some observed terrestrial isotopic ratios of the heavy elements agree well with the predictions of a theory which says that these elements were produced from lighter elements by a succession of neutron capture reactions. This is regarded as an indication that a neutron capture process had occurred in the terrestrial material before the Earth was formed. Furthermore, if the terrestrial material is typical of galactic material, neutron capture reactions probably played an important role in nucleosynthesis throughout the Galaxy.

One particularly useful type of information that can be obtained from terrestrial, lunar and meteoritic samples is provided by the *heavy radioactive elements*. The relative concentration of these radioactive elements and their decay products in rocks gives information about the ages of the Earth, Moon and meteorites, which can be compared with other estimates of astronomical ages obtained by a comparison of

18

theory and observation for stellar evolution, as has been described in *The Stars*. If we add to the observations of the radioactive elements some relatively simple theoretical ideas about the way in which all the heavy elements, including the radioactive elements, were produced, we can obtain an estimate of the ages of the radioactive elements themselves. This will be discussed in Chapter 8 after the theoretical ideas have been developed in Chapter 5.

One difficulty in studying the abundances of the elements in the Earth, Moon and meteorites is that these objects are far from being chemically homogeneous. A sample of rock from one part of the Earth's surface may be of very different composition from a sample from another part and the atmosphere and oceans are quite different again. In addition the Earth must have lost a large fraction of its light gases, hydrogen and helium, before it reached its present state. The atmosphere of any planet gradually escapes because the most rapidly moving particles have velocities greater than the escape velocity. The low mass particles have a higher velocity than the high mass particles, because at a given temperature the average kinetic energy of particles of all masses is the same, and hence the light gases escape most rapidly. When we compare abundances in the Earth with those in the Sun and stars, allowance must be made for this factor.

The Structure of the Earth

Although direct observations can only be made of the atmosphere, oceans and the outermost layer of the solid Earth, its structure is quite well understood. It is possible to say that the Earth can be divided into five or six regions as follows (fig. 9):

(*a*) Atmosphere.

(*b*) Oceans.

(*c*) Solid Crust. This is several miles deep and is much deeper under the continents than under the oceans. The reason for this difference in depth is not fully understood but it is probably connected with the phenomenon of *continental drift*.

(*d*) Mantle. This is an elastic solid of higher density than the crust which extends about half way to the centre of the Earth. It contains about 68·1 per cent of the mass of the Earth.

(*e*) Core. This is liquid and is, in turn, denser than the Mantle. The core, including the possible inner core, contains 31·4 per cent of the mass.

(*f*) Inner Core. There may exist a very small inner solid core.

It can be seen that the crust, atmosphere and oceans, which are accessible to direct observation, contain only about 0·5 per cent of the mass.

19

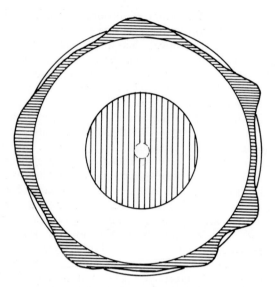

Fig. 9. The structure of the Earth. The vertically striped region is the liquid core and a possible small inner solid core is indicated. Outside the core is the mantle. The horizontally striped region is the crust above which are the oceans. (The diagram is not to scale.)

Earthquake Waves

The main sources of information about the deep interior of the Earth, which is summarized above, are earthquake waves and the Earth's magnetic field. There are several types of earthquake wave, which propagate at different speeds through different media and which are reflected or refracted at points where the physical properties change abruptly. In a solid, such as the crust or mantle, both *longitudinal pressure waves*, in which material vibrates in the direction of the waves, and *transverse shear waves*, in which material moves at right angles to the wave, can be propagated but in the liquid core only pressure waves are possible. Waves from a single earthquake can be observed at many seismological stations in different parts of the Earth, as is indicated in fig. 10, and a comparison of all the travel times enables considerable information to be obtained about the variation with depth of the properties of the terrestrial material, by using the relation between the speed of earthquake waves and the density and elastic properties of a material.

The Earth's Magnetic Field

The magnetic field of the Earth can be shown, by a study of its variation over the Earth's surface, to originate in the deep interior of the Earth. The magnitude and direction of the field is not constant in

20

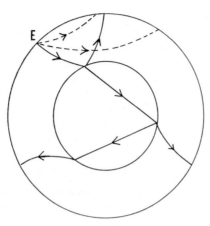

Fig. 10. Earthquake waves. The propagation of waves from an earthquake at E is shown. The dashed lines denote waves not entering the core. For the solid lines, several reflections and refractions at the core-mantle interface are shown.

time, and the magnetic North pole is found to wander with respect to the geographical North pole. Studies of the direction of magnetization of rocks of different ages in the Earth's crust show that the magnetic field of the Earth has also *reversed* completely many times in the past. When the magnetic field was first studied, it was perhaps natural to assume that it was produced by ferromagnetic material in the interior, so that the Earth behaved like a giant bar magnet. On such a model it would be difficult to understand the reversals in polarity. Now that so much is known about the field changes and it is realised that the core is liquid, it is believed that the magnetic field is produced by currents circulating in the electrically conducting liquid core. This core has a very high density, comparable to that of iron, and there has been considerable argument about whether it is largely composed of liquid metal (the iron core model) or whether, at high pressures near the centre of the Earth, non-metallic but electrically conducting material similar to that contained in the mantle is compressed to the density of iron. Throughout the Earth's history the central regions have been kept hotter than the surface by the energy released by the decay of radioactive nuclei.

Chemical Separation of Elements

Any attempt to estimate the overall chemical composition of the Earth, from the information which can be obtained directly, is closely related to the ideas concerning the past thermal history of the Earth. If, at an early stage before the crust solidified, the Earth was molten throughout, conditions would resemble those in a furnace. In a furnace it is observed that particular groups of elements

21

congregate together. Thus, if ore is melted in a furnace and is subsequently allowed to cool again, three distinct types (or phases) of material are segregated. In addition the gases which escape constitute a fourth phase. There is one type which is basically metallic, one which is mainly made of silicates and one which is concentrated in sulphur compounds. Thus, on the basis of chemical affinities observed in blast furnaces, the elements can be divided into four classes (which are not completely exclusive as some elements appear in more than one class). The groups are:

(a) *siderophile*—iron phase—elements with a tendency to associate with iron such as cobalt, nickel, gold and palladium;

(b) *lithophile*—silicate phase—elements with a strong affinity for oxygen such as lithium, sodium, potassium, magnesium and silicon;

(c) *chalcophile*—sulphide phase—elements with a strong affinity for sulphur including copper, zinc, mercury, lead and bismuth;

(d) *atmophile*—elements forming the gaseous atmosphere.

In the furnace the iron phase elements tend to fall to the bottom and the lithophile elements rise to the top in the gravitational field.

The past thermal history of the Earth and the other planets is not fully understood. Most workers now believe that the planets are composed of material lost by the Sun during its formation but that the planets did not necessarily originate as hot spheres of gas. Instead it is believed that small solid particles formed in the *solar nebula* and subsequently aggregated to produce planets such as the Earth, which may originally have been completely solid. The energy released in the decay of radioactive elements, which was much more important $4 \cdot 5 \times 10^9$ years ago than it is now, subsequently caused at least a considerable fraction of the Earth to become molten, as the core still is. Separation similar to that observed in a furnace could then have occurred giving a tendency for silicates to be concentrated in the crust and the iron phase in the core. This is in agreement with the concentration of silicate rocks in the Earth's crust and the probable existence of an iron core.

Although the tendency for some significant segregation of elements to have occurred is readily understood, it is much more difficult to decide what is the actual overall chemical composition of the Earth. Even if the complete dependence of density and elastic properties on depth could be obtained from the seismic data, there would not be a unique solution for the chemical composition as several substances can have the same density and elastic properties. Ideas concerned with separation of elements described above may make one solution seem more plausible than another. Actual estimates will be quoted later in the chapter. In converting the present composition of the Earth to its original composition, allowance must, as mentioned earlier, be made for the escape of volatile light elements.

The Meteorites

The meteorites are of several types generally known as *iron, stone* and *stony iron*, the first type being mainly metallic, the second type mainly rocks and the third type a mixture. The meteorites which are picked up on Earth may prove to be a very unrepresentative sample of the total chemical composition of the meteorites. All meteorites lose mass from heating on passing through the Earth's atmosphere and the fragment that reaches the Earth in recognizable form may be only a small part of the original meteorite. Very small meteorites will be vaporized completely. The irons suffer less than the stony meteorites from this heating in the Earth's atmosphere, because their heat of vaporization is higher. Thus the number of irons relative to stones found on the Earth's surface is not representative of the number outside the Earth's atmosphere. This is accentuated because weathering after arrival on Earth is also worse for the stones. The relative number of irons amongst meteorites picked up directly after they have fallen will be less than in meteorites discovered long after their fall and both of these observations should overestimate the number of irons.

The iron meteorites are almost purely metallic with a typical composition containing 90 per cent iron and 9 per cent nickel, although the ratio of iron to nickel shows considerable variation as different meteorites contain different alloys of iron and nickel. The stony-iron meteorites are mixtures with about equal *volumes* of stony and nickel-iron components. The stony meteorites are mainly made of silicate materials but they also contain smaller amounts of metal and sulphides and other elements, which are also of low abundance on Earth. When allowance is made for heating in the atmosphere and weathering on Earth, it appears that the most important sub-group of meteorites is one known as the *carbonaceous chondrites*. It has been estimated that about 90 per cent of all meteorites reaching the Earth are of this type. Chondrites are stony meteorites which contain small spheroidal bodies known as chondrules embedded in the stones and also contain more carbon and more water molecules than other meteorites. It is generally believed that the stony meteorites and particularly the carbonaceous chondrites give the best overall representation of the meteoritic material and abundances derived from a study of them are tabulated later in this chapter.

The Moon

Only a very small fraction of the *surface* of the Moon has yet been studied directly and it would, therefore, be surprising if a good estimate could be made of the chemical composition of the whole Moon. It is, however, already possible to draw some tentative conclusions. As successive Apollo missions are visiting different parts of the Moon's surface, the amount of information available will soon be much greater. The seismic properties of the Moon have been studied by recording its reaction to the impact of redundant stages of Apollo spacecraft, and it has been discovered that the Moon has very different seismic properties from the Earth. The most unexpected result was that the Moon oscillated for a considerable time with little damping following such an impact, whereas similar oscillations on Earth are rapidly damped. These studies indicate that the Moon does not have a sharp transition, close to its surface, comparable to the Earth's core/mantle interface. The ages of lunar samples have been estimated, as will be described in Chapter 8, and several have been found to be about $3 \cdot 5 \times 10^9$ years old. There is also an indication of the presence of older material with an age near to $4 \cdot 5 \times 10^9$ years, which is also the estimated age of the meteorites. If the Moon has the latter age, the fact that some rocks solidified 10^9 years later shows that the Moon must have had some considerable geological (or more correctly selenological) history. The report that a fragment of metamorphic rock was collected on the Apollo 15 mission supports this view. It seems almost certain, from the records of seismic detectors placed on the Moon's surface, that there is some residual volcanic activity occurring on the Moon today.

Lunar Abundances

The chemical composition of the lunar samples studied so far is notable for a lack of volatile elements and for a complete absence of water inclusions in the rocks. If the lunar abundances are compared with those in carbonaceous chondrites, it is found that the lithophile elements (see definition on page 22) are relatively more abundant in the lunar samples and that the chalcophile and siderophile elements and particularly volatile elements, such as carbon, nitrogen, sulphur and chlorine, are much less abundant. The seismic data suggest that the Moon is chemically homogeneous to a considerable depth. From the lunar samples studied so far it appears that there are definite differences between the chemical composition of lunar rocks and the Earth's crust; one particular result which was found in the earliest samples studied was a considerable overabundance of titanium. It is of interest to note that the mean density of the Moon ($3 \cdot 34 \times 10^3$ kg m^{-3}) is very close to the mean density of the Earth's mantle and it has been suggested that a major constituent in both of them could be the mineral olivene $(Mg, Fe)_2SiO_4$ which has the same density and which is a common constituent of terrestrial rocks, lunar samples and stony meteorites.

24

It is possible that the abundance differences between the lunar samples and the Earth's crust and meteorites, which have so far been found, are characteristic of the entire Moon but, even if they are, they do not seem at present to be sufficient to cause a revision of the view that the Earth, Moon and meteorites, and indeed the Sun, have a common origin. Rather, the overall similarity combined with the detailed differences can be used to try to deduce the conditions in the primitive solar nebula in which the planets are believed to have formed and to try to discover how and when the local differences in chemical composition arose. As mentioned above, one view is that the composition of the whole Moon may be generally similar to that of the Earth's mantle and the stony meteorites, so that the Moon is deficient in iron compared to the Earth. On that view, if the Moon was formed by fission of the Earth as has often been suggested, the iron core must have already separated out from the mantle material before the proto-Moon broke away.

Other Planets

Our knowledge of the chemical composition of the other planets is very slight. We do, of course, know their masses and radii and hence their mean densities. We know that Jupiter, like the Earth, has a magnetic field and that some other planets definitely do not possess magnetic fields, unless they are very much weaker than the Earth's field. In addition, the surface features of Mars have been well studied by Mariner spacecraft. The planets are usually divided into two groups: *terrestrial planets*, Earth, Venus, Mars and Mercury and *major planets*, Jupiter, Saturn, Uranus and Neptune. Pluto is very difficult to classify. In mass and radius it is nearer to the terrestrial planets than the major planets, but these properties are still not known very accurately. In space it is the most distant planet from the Sun but its orbit is also the most eccentric, which means that it is difficult to decide where it originated.

The Terrestrial Planets

The other terrestrial planets are believed to be quite similar to the Earth and the Moon, although the properties of Mercury appear to be somewhat anomalous. The masses and mean densities of the terrestrial planets are shown in Table 2. It can be seen that the mean density decreases with decreasing mass apart from the anomalously

Planet	Earth	Venus	Mars	Mercury	Moon
Mass	5·974	4·872	0·639	0·330	0·073
Mean density	5·51	5·24	3·90	5·47	3·34

Table 2. Masses (in 10^{24} kg) and mean densities (in 10^3 kg m^{-3}) of the terrestrial planets.

high density of Mercury.   There is at present no complete agreement about how the terrestrial planets were formed, although there have been suggestions that there may have been only two parent bodies, one forming the Earth, Moon and Mars and the other forming Venus and Mercury.   It is not clear how fission of such parent bodies could have produced planets in stable orbits but, if this is what happened, Mercury must have acquired a larger proportionate share of the iron than Venus and the fission in the two cases must have been quite different.

The Major Planets

Calculations have been made of the possible internal structure of the major planets, particularly Jupiter and Saturn, and it appears that the observed properties of Jupiter can be explained if its chemical composition is almost entirely hydrogen and helium.   This would imply that it has a chemical composition close to that of a typical star rather than to that of the Earth.   In fact, Jupiter probably has a greater amount of helium relative to hydrogen than the Sun and this would be consistent with its being composed of essentially solar system material with some loss of the lightest element, hydrogen, and it further supports the view of a common origin for the Sun, Earth and planets.   Although Jupiter is believed to be largely composed of hydrogen, its properties are very different from hydrogen on Earth. In the interior of Jupiter the hydrogen is under very high pressure and it is thought that it behaves like a metal, the metallic phase of hydrogen; the nuclei form a lattice and many of the electrons are free to move through the lattice like conduction electrons in a metal.

Isotopic Abundances

We have already mentioned that isotopic as well as elemental abundances can be observed in solar system material and it is notable that relative abundances of the isotopes of an element agree not only from place to place on the Earth but also very closely with those found in meteorites and lunar samples.   This is not true for isotopes which have been affected by radioactive decay processes after their formation, as will be discussed in Chapter 8.   Chemical processes do not separate the different isotopes of an element to any significant extent.   Thus, whatever chemical reactions occur in various parts of material which had a common origin, the relative isotopic abundances of any element should be essentially unaltered throughout the material. The constancy of isotopic ratios in the solar system material, which has been studied so far, gives some further support for the idea of a common origin, although it is *possible* that relative isotopic abundances are broadly similar throughout the Universe, since we have essentially no knowledge of them.

26

The main reason why the uniformity of the solar system isotopic ratios is probably significant is as follows. In Chapter 5 we shall discuss the processes whereby it is believed that element production occurs and there we shall see that different isotopes of some elements are thought to be produced by distinct processes. If nucleosynthesis occurs in stars and the distinct processes occur in a variety of types of star, there is a strong possibility that one process will have had a greater influence on the chemical composition of material in one part of the Galaxy or Universe and that another process could have been more important in a different location. Thus isotopic abundances may be very variable in the Universe and therefore uniformity of the solar system isotopic ratios may be really significant. We can, of course, use the observed isotopic ratios to try to deduce the mixture of nuclear processes which has been responsible for the production of the solar system material.

Observed Abundances

We conclude this chapter by tabulating element abundances obtained by a study of the Earth and meteorites. These are shown in Table 3. The accuracy of the numbers in this table must not be exaggerated. As has been stressed earlier, both terrestrial and meteoritic samples are not entirely representative and there is difficulty in obtaining an accurate composition of the whole Earth or of the entire group of meteorites. Further discussion of the information contained in Table 3 will be deferred until the composition of the Sun has been studied in the following chapter.

Element	log abundance	Element	log abundance	Element	log abundance
3 Li	1·7	23 V	2·5	42 Mo	0·4
4 Be	−0·1	24 Cr	4·1	44 Ru	0·2
5 B	0·8	25 Mn	4·0	45 Rh	−0·4
11 Na	4·8	26 Fe	6·0	46 Pd	0·6
12 Mg	6·0	27 Co	3·4	47 Ag	0·0
13 Al	5·0	28 Ni	4·7	48 Cd	0·4
14 Si	6·0	29 Cu	2·8	49 In	−0·7
15 P	4·1	30 Zn	3·0	50 Sn	0·3
16 S	5·7	31 Ga	1·7	51 Sb	−0·4
17 Cl	3·3	32 Ge	2·2	56 Ba	0·7
19 K	3·6	37 Rb	0·8	57 La	−0·4
20 Ca	4·9	38 Sr	1·4	63 Eu	−0·1
21 Sc	1·6	39 Y	0·7	70 Yb	−0·7
22 Ti	3·4	40 Zr	1·5	82 Pb	0·2

Table 3. Elemental abundances in the Earth and meteorites. The logarithm to base 10 of the number of atoms of each element is shown on a scale in which the number of silicon atoms is 10^6. Not all elements are shown and, in particular, those elements whose abundances are believed to be completely unreliable are omitted. The atomic number of each element is shown beside its symbol.

Summary of Chapter 2

Direct chemical analysis can be made of terrestrial rocks, meteoritic fragments picked up on Earth and of the lunar samples which have been collected by U.S. and Soviet space missions. Additional information about the interior of the Earth can be obtained by a study of the properties of earthquake waves and of the Earth's magnetic field. The Earth's central core appears to be relatively rich in iron and associated metals, whilst its outer mantle is believed to have a composition broadly similar to that of the lunar surface and the main type of stony meteorites. It is notable that the relative abundances of the isotopes of any element, which is not affected by radioactive decay processes, are the same in terrestrial, lunar and meteoritic samples. All of these results support the view that the Earth, Moon and meteorites had a common origin. The major planets, such as Jupiter, appear to have a composition much closer to that of the Sun and it is believed that the terrestrial planets have a solar composition apart from the loss of most of the light gases.

chemical composition of the sun, stars and gaseous nebulae

General principles

THE presence of a chemical element in a star is deduced from the occurrence of one or more of the characteristic *spectral lines* of that element in the radiation from the star. These spectral lines may be either *absorption lines*, when the intensity of radiation is reduced in a particular frequency (or wavelength) region (fig. 11) or *emission lines*, when the intensity is increased (fig. 12). In the radiation from most stars, absorption lines are much more common than emission lines and most element abundances are derived from a study of absorption lines and we shall discuss only these in detail, after a brief mention of emission lines.

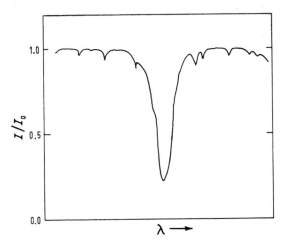

Fig. 11. An intense absorption line. The intensity of radiation is reduced to a small fraction of the intensity, I_0, at neighbouring wavelengths.

Emission lines

Emission lines usually arise when a hotter region is outside a cooler region. This sometimes happens in the outer regions of stars although the temperature does not have all the properties of a true temperature as there are important departures from *thermodynamic equilibrium.**

* The concept of thermodynamic equilibrium is described in the Appendix on page 162.

Because the temperature does not determine all of the properties of the material, it proves difficult to deduce reliable abundances from the strength of the emission lines. Some stars have shells of matter surrounding them and expanding away from them. The existence of these expanding shells can be deduced because the expanding material often gives out emission lines of the same elements which have absorption lines in the main spectrum of the star. Because of the Doppler effect, the emission lines have a different frequency from the absorption lines as explained in fig. 13. The displacement of the absorption and emission features gives the expansion velocity of the shell. We shall find that this property is important when we discuss exploding stars such as supernovae.

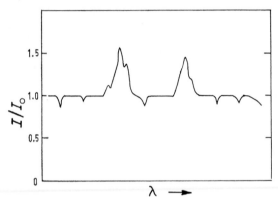

Fig. 12. Two intense emission lines. The intensity in the emission lines is much greater than that at neighbouring wavelengths. Some weak absorption lines are also shown.

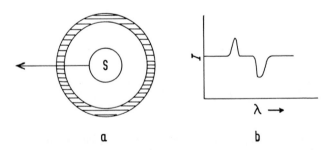

Fig. 13. Spectral lines for a star surrounded by an expanding gaseous shell. (a) Shows the geometry of the situation. A star (S) is surrounded by a gaseous shell shown hatched and the arrow points to the observer. (b) Shows a section of the spectrum. There is an absorption line from the star and, at slightly shorter wavelength, an emission line from the gaseous shell.

Absorption lines

The reason for the occurrence of absorption lines is roughly as follows. Deep down inside a star the distribution of radiation with frequency very closely follows the *black body* Planck law

$$I_\nu = B_\nu(T) \equiv \frac{2h\nu^3}{c^2} \frac{1}{\exp{(h\nu/kT)} - 1}, \tag{3.1}$$

where I_ν is the amount of radiant energy crossing unit area, in a unit solid angle about the direction normal to the area, in unit frequency range and in unit time. In expression (3.1), T is the temperature, ν is the frequency, c is the velocity of light (3×10^8 m s^{-1}), h is Planck's constant ($6 \cdot 6 \times 10^{-34}$ J s) and k is Boltzmann's constant ($1 \cdot 38 \times 10^{-23}$ J K^{-1}) and B_ν is called the Planck distribution at temperature T K. Several Planck curves are shown in fig. 14, where it can be seen that at any frequency ν the Planck function increases with temperature T. Inside a star photons emitted at any point are absorbed again at a very short distance from that point. As the surface of the star is approached, it becomes increasingly likely that a photon emitted in the outward direction will escape from the star without further absorption; at the same time the distribution of radiation with frequency begins to depart from the Planck law. If photons of all frequencies escaped from the same level in the atmosphere of the star, the radiation that we receive from the star would approximate to a black body distribution corresponding to the appropriate temperature. If, however, the

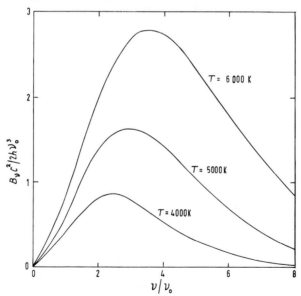

Fig. 14.   Planck curves for three values of the temperature.   The normalizing frequency ν_0 is 10^{14} s^{-1}.

31

material in the stellar atmosphere absorbs radiation of some frequencies particularly strongly, radiation at these frequencies will only escape from higher in the atmosphere, where the temperature and hence the intensity of radiation are lower. There will, therefore, be a dip in the spectrum at that frequency and an absorption line will result as shown in fig. 11.

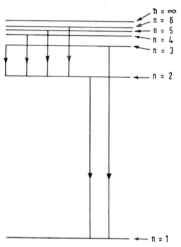

Fig. 15. The energy levels of the hydrogen atom. The vertical coordinate represents the energy difference between excited states and the ground ($n=1$) state. The arrowed transitions to the $n=2$ and $n=1$ states represent respectively emissions of the Balmer and Lyman series.

Several factors determine the presence or absence of absorption lines of a particular element in the region of the spectrum which is being observed. In the first place, the element may not possess any spectral lines in the region of the spectrum that can be observed. As an example of this, there has been no positive identification of boron in stars, although we have no reason to expect it to be absent. Secondly the element may not be present in the correct states of ionization and excitation to produce the absorption lines. At low temperatures atoms will be in their ground states and absorption lines arising from excited states will not occur. At higher temperatures, enough atoms will be in excited states for these spectral lines to be present. At very high temperatures atoms will be ionized and no absorption lines of the unionized atom can be produced.

Consider, for example, the case of hydrogen, some of whose energy levels and emission lines are shown in fig. 15. The *Balmer** lines

* Series of spectral lines in hydrogen were identified and given the names *Lyman, Balmer, Paschen, Brackett* . . . before the structure of the hydrogen atom and the existence of discrete energy levels was understood. It was subsequently realised that the four series named arose from transitions downward to the ground state and the first three excited states.

which, in absorption involve transitions from the first excited state, fall in the visible region of the spectrum, whereas the *Lyman* lines, which involve transitions from the ground state, lie in the ultraviolet. These ultraviolet lines cannot be observed with ground based telescopes, because ultraviolet radiation is absorbed strongly by the Earth's atmosphere, but they can be studied with apparatus carried in rockets and satellites; the ability to put telescopes outside the Earth's atmosphere has greatly increased our knowledge of the short wavelength radiation from stars. Some of the frequencies and equivalent wavelengths of Lyman and Balmer lines are shown in Table 4. In stars of low surface temperature, the hydrogen is all in the ground state and there are no Balmer lines. In stars of slightly higher surface temperature, the Balmer lines are very strong but in even hotter stars the hydrogen is all ionized and there are again no Balmer lines. The variation of the strength of Balmer lines with surface temperature is illustrated in fig. 16.

Lyman	α	β	γ
Frequency	24·7	29·2	30·8
Wavelength	1216	1026	973
Balmer	α	β	γ
Frequency	4·71	6·17	6·91
Wavelength	6363	4861	4340

Table 4. The frequencies (in 10^{14} Hz) and wavelengths (in 10^{-10} m $= 1$ Å) of the first three lines of the Lyman and Balmer series of hydrogen.

Spectral classification

When the spectra of stars were first studied it was thought that the differences in the spectra were essentially caused by variation of chemical composition. Subsequently, it was realised that the character of the spectrum was largely determined by the surface temperature of the star and that differences in chemical composition were relatively slight. Stars had been divided into classes known as *spectral types*, which were based on which elements were most common in the spectra. What was originally thought to be a division in terms of chemical composition is now realised to be essentially a division in terms of surface temperature. In the Harvard classification the spectral types were denoted by the capital letters A, B, C, ... It was subsequently realised that some of the groups were superfluous and that an order in decreasing surface temperature for those classes that remained was OBAFGKMRNS*. The main characteristics of the spectral types and approximate surface temperatures are shown

---

* The usual way of remembering the present order is through the mnemonic 'Oh be a fine girl kiss me right now, sweetheart'.

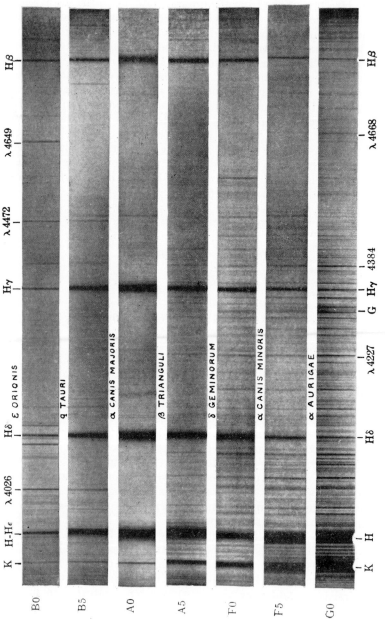

Fig. 16. The variation of stellar spectra with spectral type. Note how the strength of the Balmer lines (Hβ, Hγ, Hδ) first increases and then decreases between spectral class B0 and G0. The relation between spectral class and surface temperature is shown in Table 5. (*Photograph from Hale Observatories.*)

O	Ionized helium and metals, weak hydrogen	5×10^4
B	Neutral helium, ionized metals, hydrogen stronger	2×10^4
A	Balmer lines of hydrogen dominate, singly ionized metals	10^4
F	Hydrogen weaker, neutral and singly ionized metals	7.5×10^3
G	Singly ionized calcium most prominent, hydrogen weaker, neutral metals	6×10^3
K	Neutral metals, molecular bands appearing	5×10^3
M	Titanium oxide dominant, neutral metals	3.5×10^3
R, N	CN, CH, neutral metals	3×10^3
S	Zirconium oxide, neutral metals	3×10^3

Table 5.   Main features in the spectrum and approximate surface temperatures (in K) of stars of different spectral types

in Table 5. The simple temperature sequence ends at spectral type K. At low surface temperatures chemical composition does affect the classification. Molecules are present in the atmospheres of cool stars and which molecules are present depends critically on the exact chemical composition of the atmosphere. The spectral classes are subdivided decimally so that, for example, the hottest B-type star has class B0 and the coolest has class B9. The spectral types of some well-known stars are shown in Table 6.

Star	Rigel	Sirius	Procyon	Sun	Aldebaran	Betelgeuse
Spectral type	B8	A1	F5	G2	K5	M2

Table 6.   Spectral types of some well-known naked eye stars

Given that an element is present in the right level of ionization and excitation to allow for the existence of absorption lines in the region of the spectrum being studied, there are still two further factors determining the strength of a spectral line. One is obviously the total amount of the element present, which is what we are hoping to deduce. The other is the factor that measures the likelihood that an atom will actually absorb radiation of the correct frequency if it is present; the *cross-section* for absorption* of a photon by an atom or an ion. This factor, which is known as the *f-value* for the transition concerned, must either be calculated or measured in a laboratory. Both of these procedures are difficult. A theoretical discussion involves very difficult quantum mechanical calculations for many electron atoms. In the laboratory it is very difficult to obtain well enough controlled conditions of temperature, density and chemical composition in an

* A quantity called the cross-section can be defined for every type of atomic or nuclear reaction. In this case, if a beam of photons is incident on material which can absorb them, the number of photons absorbed will be the same as if a perfectly absorbing obstacle of some cross-sectional area A had been placed in the beam. A divided by the total number of absorbing atoms is then the effective area offered to the beam by each atom or the *absorption cross-section*.

arc or a furnace so that the number of potentially absorbing atoms is known accurately. For these reasons, many f-values are uncertain and these uncertainties must be reflected in quoted values for element abundances. Although absolute values of abundances can only be obtained if f-values are known, it is fortunate that the *relative* abundances in two stars with similar surface temperatures can be studied without knowing either the f-values or the absolute abundances; the f-value enters in the same way in the determination of each abundance and it cancels in the ratio.

Spectral line broadening

If we consider a single absorption line, we can compare the intensity of radiation in the central frequency of the line with that in frequencies just outside the line (see fig. 17). The difference between the intensity just outside the line and that at the line centre divided by the intensity outside can be called the *depth* of the absorption line. Once an f-value is known, it might be thought that the depth of an

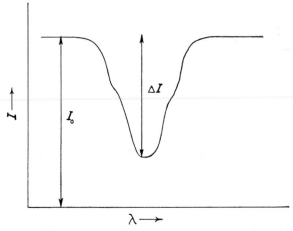

Fig. 17. The profile of an absorption line. The central depth is $\triangle I/I_0$.

absorption line would be determined purely by the number of absorbing atoms in the atmosphere. This is an oversimplification for several reasons. So far it has been assumed that a spectral line occurs at a precise frequency given by

$$E_2 - E_1 = h\nu \qquad (3.2)$$

where E_2 and E_1 are the energies of the two atomic states and the absorption line is produced by a transition from state 1 to state 2. Actually the atomic energy levels are not precise. *Heisenberg's*

36

principle of uncertainty states that in any measurement of energy there must be an uncertainty in the measurement, δE, and an uncertainty in the time of measurement, δt, related by

$$\delta E \delta t \geqslant h/4\pi \qquad (3.3)$$

As excited states of atoms have only a finite lifetime before a spontaneous transition occurs to a lower state, the energy levels have an uncertainty given by (3.3) with the lifetime of the level replacing δt. In any transition leading to a spectral line, both energy levels will have uncertainties (unless one is the ground state) and this leads to what is called *natural broadening* of spectral lines; instead of photons of only one frequency being absorbed by an atom, absorption of radiation occurs over a small, but easily observable range of frequencies.

There are several other reasons why spectral lines are broadened and they tend to lead to a much greater broadening than natural broadening. One of them arises because of the Doppler effect. If the atoms emitting or absorbing radiation are not at rest, there is either a red shift or a blue shift in their spectral lines, depending on whether the atoms are moving away from or towards an observer. In a stellar atmosphere the particles are moving about with their ordinary thermal motions. There are particles moving in all directions and typical speeds are of order $(kT/m)^{\frac{1}{2}}$, where m is the particle mass. Different atoms then absorb radiation of different frequencies and this results in the *thermal Doppler broadening* of a spectral line. There can be additional types of Doppler broadening. When we study the light from any star other than the Sun, we cannot separate the light from different parts of the star's surface. If the star is rotating, one hemisphere of the star will be moving towards the observer and the other hemisphere will be moving away, unless the axis of rotation points directly towards the observer. As a result, the Doppler effect will produce a broadening of spectral lines.

Profile of a spectral line
The radiation emitted by a star may have a relatively smooth distribution with frequency apart from a number of sharp absorption lines. This smooth distribution is known as the continuous spectrum or the *continuum*. In some stars there really are few spectral lines and the continuum is easily defined and sometimes the continuum will approximate to a Planck distribution (3.1) at some temperature. In many cases there are so many lines which are close together in frequency that it is difficult to define the continuum. For frequencies near the central frequency of a spectral line, the continuum will be nearly constant and we can consider the absorption line as a depression in the continuum. The profile of an absorption line (fig. 17), if it can be measured, should give information about the broadening mechanisms as well as the number of absorbing atoms.

The deduction of the abundance of any chemical element in principle involves the study of the profile of a large number of spectral lines. From these profiles it is possible to deduce the temperature and density of the layer in the star's atmosphere in which the spectral lines are produced and in addition to obtain the abundance of the element. Because this is a complicated process and because of the uncertainties in f-values mentioned earlier, it is rarely possible to deduce abundances to better than a factor of two and in other cases the uncertainty may be as large as a factor of ten. Even in the case of the Sun, there are uncertainties in the abundances of the heavy elements relative to hydrogen. These uncertainties are mainly due to imperfectly known f-values but they are partly caused by difficulties in calculating the structure of the solar atmosphere; for a further discussion of this latter point see page 41.

Equivalent width of spectral lines

We have spoken above as if the detailed profile of a spectral line must be studied if elemental abundances are to be deduced. In practice the measurements of a very large number of line profiles is a very lengthy task and for narrow lines it is also unreliable, because these lines suffer a further type of broadening known as *instrumental broadening*. However sharp the line is when the light leaves the star, it will be broadened in the instrument used to study it as no instrument has perfect resolution. What can be measured with reasonable ease is the *equivalent width* (W) of a spectral line (see fig. 18); the equivalent width is the width of the *continuum* which contains the same total energy as that removed in the line. The equivalent width is not altered by the instrumental profile and it can be measured accurately if the lines in the star's spectrum are not too close together, because

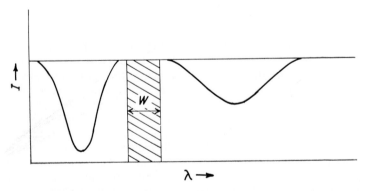

Fig. 18. The equivalent width of a spectral line. The striped area of width W has the same area as is removed from the continuum by either of the spectral lines, with very different profiles, which are shown.

38

the measurement of an area is much easier than the measurement of the line profile. In fig. 18 we show two spectral lines which have the same equivalent width; it is clear that their profiles are very different so that there is a large amount of information contained in a line profile which is not conveyed by the value of the equivalent width.

We can now ask how the profile and equivalent width of a spectral line change as the number of absorbing atoms increases. We suppose that the conditions of temperature and density do not alter; the element being considered is a minor constituent of the stellar atmosphere, so that the structure of the whole atmosphere of the star is not altered by changing the one elemental abundance. The probability that an element will absorb radiation of frequency v is expressed by an absorption coefficient $a(v)$. For any spectral line, $a(v)$ will have a maximum near the natural frequency v_0 of the line but, as a result of all the mechanisms of line broadening, $a(v)$ will be non-zero for a considerable frequency range on either side of v_0. In the case of thermal Doppler broadening the expression for $a(v)$ is

$$a(v) = a_1 \exp\left(-\alpha(v-v_0)^2\right), \tag{3.4}$$

while in the case of natural broadening it is

$$a(v) = a_2/((v-v_0)^2 + \beta^2), \tag{3.5}$$

where a_1, a_2, α and β do not depend on v. In the general case the complete absorption coefficient will involve both these types of broadening and others as well.

Curve of growth

A theoretical description of how the equivalent width, W, depends on the number of absorbing atoms, N, is as follows. When the number of atoms is small, most energy is removed from near to the line centre and, as the number of atoms increases, W increases linearly with N. As the number of atoms increases further, saturation occurs at the centre of the line; however many more absorbing atoms there are, the amount of radiation removed at the centre of the line cannot increase further. Absorption can still occur in the *wings* of the line but, as absorption is less likely away from the line centre, the equivalent width increases less rapidly than linearly with N. Using the expression for the absorption coefficient, it is possible to calculate W as a function of N and to obtain what is called the *curve of growth* for the line, which is shown in fig. 19.

Now, of course, we cannot *observe* a curve of growth in the form in which I have described it as we cannot change the proportion of an element in the composition of a star and watch how the equivalent width of a line varies. Fortunately, theory tells us that if we consider

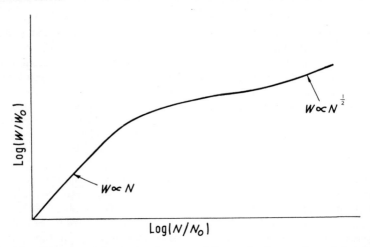

Fig. 19. The curve of growth of a spectral line. As the number of absorbing atoms, N, is increased, the equivalent width, W, of a spectral line first increases linearly with N. After a flatter region it eventually increases as $N^{\frac{1}{2}}$ (N_0 and W_0 are constants).

a series of lines of the same chemical element in the same state of ionization, which have different f-values, as not all transitions are equally probable, we can produce such a curve. The equivalent width of a line is determined by Nf so that lines of transitions with a high f-value will look just like lines in an atmosphere with a higher abundance of the element. In fact, it turns out that if we plot log (W/λ), where λ is the wavelength of the line, against log $f\lambda$ for all of the unionized iron lines in the Sun (for example), we obtain a curve of growth which is shown in fig. 20. The use of curves of growth in

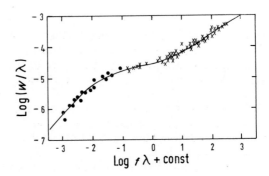

Fig. 20. The curve of growth of neutral iron in the Sun. The crosses represent different iron lines. To complete the curve lines of neutral titanium (filled circles) are also used.

obtaining relative abundances for several stars is described on page 42, after we have made some general remarks on the structure of stellar atmospheres.

Structure of stellar atmosphere

From the theory of the solar atmosphere, we *should* be able to deduce (for example) the absolute iron abundance of the Sun, but this is a complicated procedure. The theory of stellar structure described in *The Stars* explains how the total amount of radiation emitted by a star can be determined if its mass, chemical composition and age are known. Although it also determines the effective temperature, T_e, of a star, it does not enable us to discover the precise distribution with frequency of the radiation emitted by a star and how this differs from a black body distribution with temperature T_e. In *The Stars* it has been assumed that the radiation departs only slightly from the black body form and that the net outward flow of radiation arises because there is a very small difference between the outward flowing and inward flowing radiation. Near the surface of a star this is no longer true; the radiation departs significantly from the Planck form and most of the radiation is flowing outwards. To study the structure of this atmosphere, it is necessary to replace a single equation saying how radiation flows down a temperature gradient by a coupled set of equations describing the separate flow of radiation of all frequencies.

In principle, if the chemical composition of the star were known, these equations could be solved in conjunction with the other equations of stellar structure and a complete description of the star's spectrum, including the profiles of all of the absorption lines, would be obtained. In practice this cannot be done; even if all of the *f*-values for absorption and emission of radiation by the atmospheric atoms and ions were known exactly, the amount of mathematics involved would be too great even for modern computers. Fortunately, as hydrogen and helium are much more abundant than other elements in most stars, it appears that the general shape of the continuous spectrum and the variation of temperature with depth in the stellar atmosphere can be calculated without including contributions to the absorption and emission of radiation by many of the less abundant elements. These can then be included subsequently to explain how particular spectral lines are carved out of the continuum.

When the variation of temperature and density with depth in the stellar atmosphere is known, it is possible to determine where in the atmosphere the absorption lines due to an atom in a particular state of excitation and ionization will be mainly produced; thus lines due to neutral iron arise in the low temperature part of the atmosphere because at higher temperatures iron is ionized. By studying the flow of radiation at frequencies close to that of a particular spectral line in the relevant atmospheric layers, using an assumed abundance

41

of the element, the profile and equivalent width of the line can be calculated. The assumed abundance can then be adjusted until the correct equivalent width and a reasonable approximation to the correct profile is found. This gives an abundance *relative to hydrogen* for the element concerned.

Determination of relative abundances

In this manner we can determine, for example, the iron abundance of the Sun. We can do this only subject to the uncertainties in f-values and the properties of the solar atmosphere which have been mentioned earlier. What we can do with more certainty is to study the curves of growth for neutral iron for other main sequence stars of the same spectral type as the Sun; these are stars with generally similar properties to the Sun but they may have just those detailed abundance differences for which we are looking. If the stars have the same spectral type and surface temperature, any differences in the curve of growth should be due to differences in the N_{Fe} from one star to the other, since W depends on Nf and the f-values will be the same. A comparison of the curves of growth gives a reliable value for the relative iron abundances in the two stars.

Thus, reasonably reliable relative abundances can be obtained from this so-called *coarse analysis* of the spectra. The method depends on the observer's ability to recognize that two stars are basically similar but that they may have detailed abundance differences. There is one relatively quick way of obtaining a rough idea of heavy element abundances in stars. Observers frequently measure the total radiation received from stars in three bands of wavelength about 1000 Å wide called the U, B and V bands, which are situated in ultraviolet, blue and yellow regions of the spectrum, the central wavelengths of the bands being approximately

$$\lambda_U \approx 3650 \text{ Å}, \ \lambda_B \approx 4400 \text{ Å}, \ \lambda_V \approx 5480 \text{ Å}. \tag{3.6}$$

Most spectral lines occur in the U and B region of the spectrum. If a star has a low heavy element content compared to another star of the same surface temperature, its radiation in the U region will be stronger as less radiation has been removed from the continuum by lines (the radiation removed in the spectral lines is redistributed through the whole spectrum as the total energy radiated by the star is not affected by the presence of lines). Such a star is said to show an *ultraviolet excess* and this is a clue to low metal content, which puts a star on a list which is worth studying more closely.

These coarse analyses of chemical composition must be followed by a much more detailed study if the relative abundances are to be superseded by reliable absolute abundances, but they are good techniques for discovering stars which probably have unusual abundances. The detailed study of stellar atmospheres, spectral lines and abundance determinations is very complicated, as has been indicated above, and here we can do no more than describe the results obtained.

Significance of stellar abundances

As our aim is to try to decide what pattern emerges from the observed abundances, it is useful to have a preliminary discussion about what atmospheric abundances mean. There are essentially three things that the atmospheric abundances could represent. These are:

(i) the atmospheric abundance of the star at birth; that is, the initial composition of the star;

(ii) the present abundance of the interstellar medium if the star has *accreted* interstellar matter;

(iii) the initial abundance modified by nuclear reactions inside the star, if it has lost a large amount of mass or if material from the central regions, where nuclear reactions have occurred, has been carried to the surface by mixing currents.

Certainly factor (iii) is important in some very evolved stars but, if initially we discuss only stars near to the main sequence, it should not be very important. Accretion of interstellar matter occurs when it is attracted by the gravitational field of the star and is captured. Substantial accretion only occurs if the star moves slowly through a dense interstellar gas cloud and, from observations of the interstellar gas density, it appears that significant accretion is a very rare event. In contrast a wide variety of stars are observed to be losing mass; even the Sun is observed to lose between 10^{-14} and 10^{-13} of its mass each year. These observations suggest that mass loss is more important than accretion, so that the observed composition of stars that are not highly evolved, so that no significant nuclear reactions have occurred in their interiors, should be their initial composition. We shall assume that this is generally true.

As has been mentioned earlier, the differences in chemical composition between stars are very much less than appears from their spectra; the elements which are most prominent in a spectrum are often not those which are the most abundant, but are those which are most capable of absorbing radiation of the wavelength being studied. Some elements may not be detected in the atmosphere of a star because they do not possess spectral lines in the correct region of the spectrum or are not in the correct states of ionization and excitation to produce them. Thus our knowledge of stellar chemical compositions will always be imperfect.

Problem of helium abundance

A particularly important example is that of helium. In cool stars not only will helium not be ionized but it will be in its ground state. All

43

of the absorption lines of cold neutral helium lie well into the ultra-violet region of the spectrum, which means that, because of absorption in the Earth's atmosphere (see fig. 21), they cannot be observed in telescopes on the Earth's surface. The same is true of the Lyman lines of hydrogen, which involve transitions from its ground state. The Lyman lines can be observed by telescopes mounted in rockets and satellites. This is not generally true of the helium lines, because radiation with wavelengths shorter than the hydrogen Lyman lines is absorbed by the cold hydrogen which is an abundant constituent of the interstellar gas and we cannot observe that part of the spectrum of distant stars. Thus helium, even if present, cannot be seen in the majority of stars. In most stars, which are hot enough to show spectral lines of ionized helium or neutral helium in higher states of excitation, the helium abundance is high and it appears to be the only element which may have an abundance comparable with that of hydrogen. There is great interest at present in trying to determine whether all stars contain a considerable amount of helium, which may have existed when the Galaxy was formed. This will be discussed further below and at length in Chapters 6 and 7.

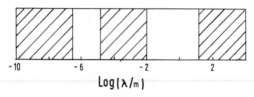

$$Log(\lambda/m)$$

Fig. 21. The transparency of the Earth's atmosphere. Electromagnetic waves of wavelengths corresponding to the hatched areas are almost completely absorbed in the Earth's atmosphere. Between the hatched areas are the visible *window* and the radio *window*.

The solar abundances

It is usual to express the abundances of elements in other stars in terms of solar abundances, which should be amongst the most reliable, and we start by tabulating some recent evaluations of the solar abundances in Table 7. Not all elements are shown; some are absent because a selection has been made and others are absent because they cannot be observed or because the theory of stellar atmospheres does not yield a reliable abundance. The latter is true of helium. Although it was discovered in the Sun before it was found on Earth, it is not possible by direct means to obtain a good estimate of its abundance. The observed helium lines are formed in the high solar atmosphere. The solar atmosphere is peculiar in that it has a temperature minimum just above the *photosphere,* or visible surface and then the temperature (measured by average kinetic energy of the particles) rises to a very high value in the *chromosphere* and *corona*. Helium is observed in emission lines from the chromosphere but conditions are so far from thermodynamic equilibrium (see Appendix on page 162) that reliable abundances cannot be obtained.

It should be remembered that all values in Table 7 are uncertain and the uncertainties can be of order a factor of 2 or equivalently of order 0·3 in the logarithm; because of the habit of expressing abundances in logarithmic terms, the errors tend to appear smaller than they are. The present values cannot be regarded as final values. In the past few years new measurements or calculations of f-values have led to significant revisions in the accepted abundances of some of the more abundant elements such as carbon, nitrogen, oxygen and iron. The latter revision involved a change nearer to a factor of ten than a factor of two.

Element	log abundance	Element	log abundance	Element	log abundance
1 H	10·5	21 Sc	1·3	41 Nb	0·5
3 Li	−0·7	22 Ti	3·2	42 Mo	0·4
4 Be	−0·4	23 V	2·2	44 Ru	−0·1
5 B	<1·3	24 Cr	3·9	45 Rh	−0·7
6 C	7·1	25 Mn	3·4	46 Pd	−0·3
7 N	6·4	26 Fe	6·1	47 Ag	−1·4
8 O	7·3	27 Co	3·1	48 Cd	0·5
11 Na	4·7	28 Ni	4·7	49 In	0·2
12 Mg	6·0	29 Cu	2·7	50 Sn	0·2
13 Al	4·9	30 Zn	2·9	51 Sb	0·4
14 Si	6·0	31 Ga	1·4	56 Ba	0·5
15 P	3·9	32 Ge	1·8	57 La	0·3
16 S	5·7	37 Rb	1·1	63 Eu	−1·0
17 Cl	4·1	38 Sr	1·3	70 Yb	0·0
19 K	3·5	39 Y	0·8	82 Pb	0·3
20 Ca	4·8	40 Zr	0·7		

Table 7. Elemental abundances in the Sun. The logarithm to base 10 of the number of atoms of each element is shown on a scale in which the number of silicon atoms is 10^6. Not all elements are shown and, in particular, those elements other than boron whose abundances are believed to be completely unreliable are omitted. The atomic number of each element is shown beside its symbol.

One conclusion can be drawn immediately from these solar abundances. If we make allowances for their uncertainties, the solar abundances are in good agreement with those determined for the Earth and meteorites, which have been described in Chapter 2. There are detailed differences. Some elements cannot be observed in the Sun, and the Earth and meteorites are very deficient in volatile light elements, and some other abundances are unreliable. It is, however, quite plausible that the Sun and planets were formed out of the same material. There is a very clear discrepancy in the case of

45

the light elements lithium and beryllium, which are much more abundant in the Earth and meteorites than in the Sun. When we discuss cosmic rays in Chapter 4, we shall see that Li and Be are even more abundant in them and in Chapter 5 we shall learn that they are not produced by any of the best known processes leading to the origin of the elements. The nuclei of these light elements are not very stable and they are easily destroyed by nuclear reactions at temperatures lower than those that occur in the interiors of most stars. It is, therefore, not surprising that their abundances in the Universe are very variable.

Composition of other stars

When the chemical compositions of other stars is studied, we can divide the stars into three main classes:

(i) stars with abundances like the Sun; e.g. those stars in which the abundances of none of the elements, which can be observed in both star and Sun, differ by more than a factor of about two relative to some element taken as standard*;

(ii) stars whose composition is quite clearly different from that of the Sun but which show no pecularities in their other properties;

(iii) stars with peculiar spectra associated with other unusual properties; e.g. stars with strong magnetic fields which vary with time and stars with time-varying spectra.

We are unable to discuss the stars of type (iii) in this book, although a proper understanding of their spectra is both difficult and important. In some cases we are uncertain whether the stars do have peculiar abundances or whether other properties of the stars are producing unusual spectra when the abundances are normal. In other cases we feel sure that the abundances are unusual but that it is possible that nuclear processes in the star itself may be responsible for the unusual abundances rather than their being representative of the chemical composition of the star at birth. In many cases the stars with peculiar abundances have passed through part of their evolution, so that there has been an opportunity for composition changes.

We are interested then particularly in the stars of type (ii) and particularly in three of their properties. We wish to know in which way their abundances differ from those of the Sun, whether these abundances are correlated with the ages of the stars and whether they are correlated with position, or more precisely place of origin, in the Galaxy. If our normal ideas about the formation of stars from the

---

* There are greater differences in the case of lithium and beryllium which have just been mentioned.

interstellar gas are correct, *age** and *place of origin* should determine the chemical composition of a star. If the chemical composition of the interstellar medium has always been spatially uniform, age alone should determine the chemical composition of a star, provided no significant changes have occurred during its life history. If the interstellar medium has been appreciably non-uniform, the place of origin of the stars should also be important. Eventually, we hope that study of the composition of many stars will give knowledge of the gradual change of chemical composition of the interstellar gas with position and time, which can be combined with theoretical con- siderations of the types described in Chapters 5 and 7.

Helium content of stars

We have already said that there is considerable uncertainty concerning the abundance of helium in many stars. Those stars, for which what is thought to be a reliable helium abundance can be obtained, typically have between 25 per cent and 35 per cent of their mass in the form of helium. The solar helium abundance has been estimated in two ways. Theoretical calculations of the structure of the Sun suggest that its observed luminosity and radius can be understood if it contains between 25 and 30 per cent helium. A small fraction of the cosmic rays which will be discussed in the next chapter are produced in the Sun. The relative abundances of He, C, N and O in the solar cosmic rays and H, C, N, O in the solar atmosphere can be studied. The ratios C : N : O are the same in both cases and this enables the helium in the cosmic rays to be compared with the hydrogen deduced from the spectrum. This leads to a helium abundance which is nearer to 20 per cent than to 25 per cent but it is not certain that there is a discrepancy with the other value. From time to time it is reported that stars have been discovered which have virtually no helium but, seemingly, no other peculiarities in their spectrum: no such observation has been definitely confirmed. As we shall see in Chapter 6, there is great interest in whether any stars, including the Sun, have less than 25 per cent helium by mass, as a simple *cosmological theory* suggests that the Galaxy should have contained about that fraction of helium when it was formed.

Heavy element content of stars

When we consider the elements heavier than helium we are on somewhat firmer ground. The Sun contains about 98 per cent hydrogen and helium and only about 2 per cent in the form of heavier elements. This great preponderance of hydrogen and helium seems to be true in all *normal* stars, that is those in classes (i) and (ii) on page 46. There is no clear evidence for stars with a substantially

---

* More exactly we mean time of formation but the present age can be used instead of time of formation.

greater amount of heavy elements than the Sun. There are probably stars whose metal* content is twice that of the Sun, but much more than that is very unlikely. In contrast, there are stars with very much lower metal content than the Sun. Stars in globular clusters, for example, are all deficient in heavy elements and in some clusters the deficiency is by a factor of 100 relative to the Sun.

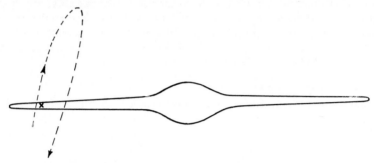

Fig. 22.   The orbit of a high velocity star in the Galaxy.   A star, which is at some time close to the Sun (×) may penetrate far into the halo.   No attempt has been made to take account of galactic rotation in this schematic diagram.

When *metal-deficient* stars were first discovered, it appeared that the metal deficiency was closely correlated with age. The oldest stars to which we can give a reasonably reliable age are those in globular clusters and these have the lowest heavy element abundances. Other stars in the solar neighbourhood which are metal-deficient are high velocity stars. These stars are at present moving in the disk of the Galaxy, but in the course of time they will penetrate far into the halo of the Galaxy (see fig. 22). This indicates that their place of origin could well have been in the halo. In trying to understand the structure and evolution of the Galaxy, it is necessary to ask why most of its mass is in the highly flattened disk but that some mass is contained in an approximately spherical halo. One of the most popular theories suggests that the Galaxy collapsed from a large radius to its present size. Much energy was dissipated by collision of interstellar gas clouds so that the material of the disk could not then re-expand to its original radius. Only that matter which had already condensed into stars could continue to move in orbits through-out the halo. On this picture, the high velocity stars are very old and the original metal content of the Galaxy is thought to have been very small.

Although the oldest *halo* stars are very metal-deficient, there is evidence that the oldest stars in the *galactic disk* have abundances

* Astronomers often loosely use the word *metal* to describe all elements other than hydrogen and helium.

which differ little from the Sun. By methods which have been discussed in detail in *The Stars*, it is possible to estimate the ages of both galactic and globular star clusters and it has been found that, although the oldest galactic clusters are only a little younger than the globular clusters, their chemical compositions are not significantly different from that of the Sun. *This has led to the suggestion that about half of the heavy elements in the Galaxy must have been produced in a short period after the origin of the Galaxy, perhaps in the first 10 per cent of its life.* There must have been a much slower rate of production since then. Those stars in which a low helium abundance has been postulated are also amongst the oldest and, *if the initial helium content of the Galaxy was low, a large fraction of the helium must also have been produced in the very early stages of galactic history.*

Variation of composition with galactic position
Originally it appeared that disk stars of the same age had essentially the same metal content but more recent observations have indicated that metal content relative to hydrogen in stars of the same age might differ by a factor of two or more. This means that place of origin as well as age is important in determining the chemical composition of a star. At any time there must have been spatial variations of the chemical composition of the interstellar gas in the Galaxy but it is too early to say in detail what these have been. There is a definite indication that there are higher metal abundances near the centre of the Galaxy than elsewhere. The two results which we have just mentioned, that there was initially a rapid burst of metal production in the Galaxy and that the interstellar medium must have been significantly spatially inhomogeneous, will be very important when we consider in Chapter 7 where the heavy elements have been produced.

If any detailed conclusions are to be reached about the past variations of chemical composition in the Galaxy, it is necessary to be able to compare stars of the same age but different places of origin. The latter point is particularly important because stars moving at only 10^4 m s^{-1} (a slow stellar speed) can travel a large distance in a few hundred million years and stars which are in the solar neighbourhood today need not have formed there. Studies of stellar evolution show how the observable properties of a star, luminosity and surface temperature, change as it evolves. For some stages in the early evolution of a star, the dependence of these properties on the star's age is such that a good value of the age can be deduced from the observations. If we know the age of a star and its present position and velocity and if we calculate the galactic gravitational field from the mass distribution in the Galaxy, we can solve for the orbit of the star backwards in time to its position of origin. In this calculation collisions with other stars can be neglected because their sizes are so much smaller than their mean separations. If this calculation is done for stars which are at present near to the Sun and if allowance is made for uncertainties both in age estimates and in the gravitational field, it appears that the stars were formed mainly in the regions of spiral arms, and that the metal contents of stars formed nearer to the centre of the Galaxy than the Sun is higher than that of stars formed further out in the Galaxy (see fig. 23). The results of this study are at present suggestive rather than conclusive and the results of further work are awaited with interest.

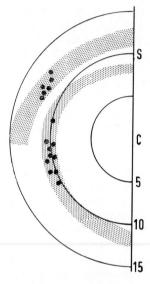

Fig. 23. The correlation of chemical composition with place of origin of stars. The filled circles represent the calculated places of origin of stars which are at present near to the Sun (S). C is the galactic centre, the figures denote the distance from C in 10^3 parsec and the stippled areas are spiral arms. The stars formed in the spiral arm further from the galactic centre than the Sun appear to have a lower metal content than those formed in the inner arm.

In a longer book space would now be devoted to a discussion of the observed composition of individual stars and in particular to a study of the detailed *mix* of the heavy elements. If one star has twice the metal content of another, is the content of all metals changed by this factor or are there important differences from element to element? There is certainly some evidence that this mix does vary from star to star but, bearing in mind the uncertainties of the abundances *the overwhelming impression is the great similarity of the element mix from star to star*, apart from variations due to nuclear processes which have occurred in the stars themselves. Unless we try to go much deeper into the subject than is possible in this book, *we are justified in treating the heavy elements as a single entity.* This will also be important when we consider the processes producing the heavy elements and the sites in which they have been produced.

Abundances in the interstellar medium and gaseous nebulae

There is a very important difference between the physical conditions in stars and in the interstellar medium. Inside a star conditions are close to those of thermodynamic equilibrium (for details see Appendix), so that all the properties of the material are essentially determined by its temperature, density and chemical composition. Near the surface of the star this is no longer completely true, but the radiation from stars bears some resemblance to the Planck function (3.1) and we can define the star's surface temperature. In contrast, in the interstellar medium the physical conditions differ widely from thermodynamic equilibrium. For example, in interstellar space the radiation from stars has a distribution with frequency approximately that of a black body of about 10^4 K but it is much less intense. Black body radiation has an energy density aT^4 J m^{-3}. The energy density in interstellar space is only that appropriate to a temperature of 3 K; the radiation is said to be *diluted* by a factor of $(10^4)^4/3^4 \approx 10^{14}$. When we refer to temperature below, it will be a *kinetic temperature* determined by the mean kinetic energy of the particles.

Because conditions in the interstellar medium do differ so greatly from thermodynamic equilibrium, it is difficult to deduce reliable element abundances. Values can be obtained which are rather less reliable than those quoted for stars. The interstellar medium can be studied in several ways. It can be divided into two parts. There are cold (~ 100 K) regions, which are known as *HI regions*, because there the hydrogen is not ionized and *HI* is the spectroscopic designation for unionized hydrogen. There are also *HII regions*, which are hotter ($\sim 10^4$ K) and in which hydrogen is ionized. The cold interstellar medium was first discovered by the existence of absorption lines in stellar spectra, which occurred at the same wavelength for all stars in the same direction and therefore did not share the Doppler shifts due to the motions of the individual stars. These could be interpreted as being produced by absorbing matter between the star and the observer. Later, interstellar hydrogen was discovered and studied because cold hydrogen both emits and absorbs radio waves at a wavelength of 21 cm. Very recently radio astronomers have discovered many other molecules such as formaldehyde and ammonia. The discovery of organic molecules has led to speculation that living material may first have originated in interstellar space.

Although the occurrence of various elements in interstellar space can be demonstrated in this manner, for most elements the best estimates of abundances come from a study of the hot gaseous nebulae which have temperatures similar to those of stellar atmospheres. Whereas in the case of stellar atmospheres abundances are determined from absorption lines, in the case of gaseous nebulae emission lines must be used. An interpretation of the strength of these emission lines in terms of element abundances is not always easy, because of the lack of a well-defined temperature governing all processes, but abundances can be deduced and it is found that the chemical composition of the hot gaseous nebulae is similar to that of very bright recently formed stars. Results for the Orion nebula and stars in its neighbourhood are shown in Table 8. These results are consistent with the idea that the young stars have recently been formed out of the interstellar gas.

Elements	H	He	C	N	O	Ne	S	Cl	Ar
Nebula	12·0	11·1	8·4	7·7	8·6	8·8	8·0	5·9	6·6
Star	12·0	11·2	8·2	7·9	8·8	8·9	7·7	6·2	6·9

Table 8. Abundances of selected elements in the Orion nebula and neighbouring B stars. The logarithm to base 10 of the number of atoms of each element is shown on a scale in which the number of hydrogen atoms is 10^{12}.

Abundances outside the galaxy

Although we are interested in the chemical composition of the Universe, little detailed information exists for objects outside our Galaxy. Stars in other galaxies are mainly too faint for their spectra to be studied in detail although the composite spectra of groups of stars can be studied and compared with groups in our Galaxy. It is also possible to make a detailed study of large gaseous nebulae in nearby galaxies. The conclusion is that the overall composition of nearby galaxies is not greatly different from our own and that we should, at least, be able to treat it as representative of our neighbourhood in the Universe. In addition recent studies of gaseous nebulae suggest that other galaxies have their highest metal contents in their central regions.

Composite abundance curve

Because of the general similarity in abundances between the Earth and meteorites, the Sun and other stars, and gas clouds in the disk of the Galaxy, in the case of those elements which can be studied in all three, it is natural to assume that the abundances of the unobservable elements are likely to be the same. It is therefore possible to produce a *composite abundance curve* from these sources which can be taken to be characteristic of the chemical composition of the galactic disk; the low abundance of volatile light gases on the Earth is, of course, disregarded. Such a composite abundance curve is shown in fig. 24, a schematic version of which has already been shown as fig. 7 in Chapter 1. While this gives a good general idea of the chemical composition of the disk of our Galaxy, the uncertainties in absolute values of abundances must again be stressed and it must be remembered that some individual objects have metal abundances that differ by a factor of at least two from those shown in fig. 24.

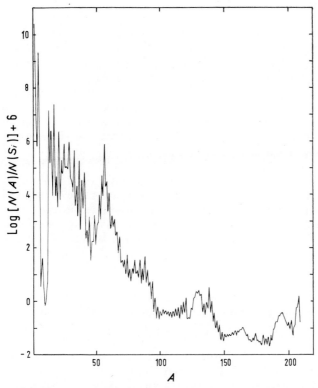

Fig. 24. The solar system abundance curve. The abundance of the element of atomic mass number A is shown on a scale on which the abundance of silicon is 10^6.

The general character of the abundances is, however, clear. Hydrogen and helium are much more abundant than the heavier elements and there is a gradual decline in abundance towards high atomic number. Superimposed on this decline to high atomic number there is some fine structure. Nuclei containing even numbers of both protons and neutrons tend to be more abundant than their neighbours and those which can be considered to be composed of α particles are particularly abundant. As mentioned earlier, the elements lithium, beryllium and boron have very low abundances. There are also some pronounced local abundance peaks. One peak is in the neighbourhood of iron where abundances are two or three orders of magnitude higher than if they lay on a smooth curve and there are also small (double) peaks near mass numbers 90, 135 and 200. We shall see in Chapter 5 that, apart from the very high abundance of hydrogen and helium, most of the peaks and other irregularities are associated with specific properties of nuclei.

Summary of Chapter 3

The presence of a chemical element in a stellar atmosphere is detected by the occurrence of its emission or absorption lines in the spectrum. Element abundances are normally determined by a study of absorption lines. There is a great variety in stellar spectra but this is mainly produced not by a large difference in chemical composition but by a difference of surface temperature. The intensity of a spectral line depends not only on the temperature, density and chemical composition of the stellar atmosphere but also on the probability that an atom will absorb a photon. This *f*-value is often difficult to measure or calculate and for this reason element abundances are often uncertain by a factor of two or more. More accurate values can be obtained of the relative abundances in two stars with similar surface temperature.

For most stars, and in particular for main sequence stars, it is believed that the atmospheric composition is similar to the original composition and hence to that of the interstellar medium out of which the star formed. Thus, a study of stars of varying age and position should enable the evolution of chemical composition of the Galaxy to be investigated. Abundances are usually expressed in terms of the solar composition, which is very similar to that of the Earth and meteorites after allowance for special factors such as the loss of volatile gases by the Earth. The oldest known stars are very metal-deficient compared with the Sun but the metal content of the Galaxy appears to have increased very rapidly shortly after they were formed. Most other stars appear to have a metal content which differs by no more than a factor of about two from that of the Sun, but there is evidence for spatial variations of metal content in the interstellar gas at a given time. It is *possible*, but not certain, that all stars have a high helium content comparable with that of the Sun.

CHAPTER 4
element abundances in cosmic rays

Introduction

Cosmic rays are high energy charged particles which appear to fill the Galaxy and which may also pervade intergalactic space. Their total *number density* (number in unit volume) is very low but, because they individually possess high energies, they make a contribution to the total energy of the *Galaxy* which is comparable, for example, to that of all the photons in interstellar space. It is not at present clear whether all intergalactic space is filled with cosmic rays to anything like the same density but, if it is, they form a very important constituent of the *Universe*. A few numbers will illustrate this. A typical cosmic ray particle has an energy of about 10^{10} eV $(1 \cdot 6 \times 10^{-9}$ J), but particles have been observed with energies up to 10^{20} eV. At 10^{10} eV a proton is moving with speed $0 \cdot 996 \ c$, while at 10^{20} eV its speed is $(1 - 4 \times 10^{-23}) \ c$. The total energy density of cosmic rays in the neighbourhood of the Earth is found to be about 10^6 eV m^{-3}, while the average density of matter is 10^6 protons m^{-3}, which by Einstein's mass-energy relation $\mathscr{E} = mc^2$ gives a *rest mass energy density* of 9×10^{14} eV m^{-3}. Thus the contribution of cosmic rays to the energy of the *Galaxy* is seen to be small. In contrast, the average density of matter throughout the *Universe* is believed to be between 10^{-1} and 10 protons m^{-3}. This gives a rest mass energy density between 9×10^7 and 9×10^9 eV m^{-3}. From these figures, it can be seen that, if cosmic rays fill the Universe with a density similar to that near the Earth, their total energy is between 10^{-4} and 10^{-2} of the total rest mass energy of matter. We shall see that this is important when we discuss their origin.

Chemical composition of cosmic rays

It is possible to study the chemical composition of the cosmic rays and to compare this with the chemical composition of the solar system and the stars. In order to be studied, cosmic ray particles must be slowed down or brought to rest in a detector. There is a variety of detectors which enable something to be learnt of both the energy and the charge of the particle being detected. Except in the case of ions with low electric charge, it is difficult to obtain an exact value for the nuclear charge. Thus it is not possible to state a precise chemical composition for the cosmic rays but an approximate division into charge ranges can be obtained and this is shown in Table 9, which also contains some stellar abundances. The properties which are immediately apparent from this table are:

(a) There are relatively more heavy elements in cosmic rays than in a typical stellar mixture. This is true of elements such as carbon and oxygen and even more so for iron.

(b) There are considerably more of the nuclei of the light elements, Li, Be and B than in a stellar mixture.

In addition, but this is not shown in the Table, there is strong evidence that there are some extremely heavy nuclei in the cosmic rays, possibly including transuranium elements.

Nuclear charge	Cosmic ray abundance	Stellar abundance	Ratio
1	1	1	1
2	5.0×10^{-2}	8.0×10^{-2}	0.6
3– 5	1.3×10^{-3}	3.0×10^{-9}	4×10^5
6– 9	4.0×10^{-3}	8.0×10^{-4}	5
10–14	1.2×10^{-3}	2.5×10^{-4}	4.8
15–19	1.3×10^{-4}	1.6×10^{-6}	80
20–25	1.3×10^{-4}	1.6×10^{-6}	80
26	3.0×10^{-4}	1.6×10^{-5}	19

Table 9. Element abundances in cosmic rays, observed at the top of the Earth's atmosphere compared with characteristic stellar abundances. The abundances by number are expressed as a fraction of the hydrogen abundance in each case.

Origin of cosmic rays

In order to interpret these results and to see what relevance they might have for the origin of the chemical elements we must discuss the *birth* and *life history* of cosmic rays. By birth we mean the process whereby the particles were given the high energies which they at present possess. In discussing the life history of the cosmic rays, we recognize that they suffer collisions with low temperature gas in the Galaxy, which has been discussed on page 51. One effect of this is that the chemical composition of the cosmic rays gradually changes; protons of the interstellar gas are moving with very high speeds relative to the cosmic ray particles and can collide with them and knock protons and α particles (for example) out of them. In the same way cosmic ray particles can break up heavy nuclei in the interstellar gas.

When the birth process of cosmic rays is considered, there are three possible ways of accounting for their chemical composition, which differs from that of stars and the interstellar gas. These are:

(i) production in a region rich in heavy elements.

(ii) preferential acceleration to high energy of heavy elements.

(iii) production of heavy elements and cosmic rays are essentially the same process.

56

The light elements Li, Be, B are not believed to have been produced in the birth process and their origin is discussed below.

Supernovae and cosmic rays

Cosmic rays have such extremely high energies that it is difficult to believe that a mechanism can be found to give them these high energies without at the same time conditions being such that some nuclear reactions are occurring simultaneously. (Even an energy of 1 eV corresponds to a typical thermal energy at 10^4 K; cosmic ray energies clearly cannot be thermal.) Thus it is believed that process (iii) above must be important and there is now a rather general belief that it is more important than (i) and (ii). If this is so, the origin of cosmic rays is closely linked with the origin of the chemical elements. To look forward to Chapter 7, it can be mentioned that it is believed that a very important source of cosmic rays is the explosion of supernovae. These are stars whose brightness suddenly increases by many orders of magnitude over its previous value to such an extent that a single supernova can give out the same amount of light as a whole galaxy of stars. Simultaneously the star is apparently shattered by an explosion and a large amount of matter is expelled into interstellar space. It is believed that a considerable amount of nucleosynthesis occurs in supernovae and, if the cosmic rays are produced at the same time, they may give information about element production. (Supernovae explosions are discussed on pages 118 to 122.)

Interpretation of the Doppler displaced spectral lines in supernovae spectra shows that the bulk of the matter expelled from supernovae travels with speeds of order $10^6 - 10^7$ m s^{-1}, which is of order $0.01\ c$. What is required for cosmic ray production is that a large fraction of the energy released in the explosion accelerates a very small fractional mass of the ejected material to a speed very close to that of light. Although we have no direct evidence that these relativistic particles are produced in supernovae, observations that supernova remnants such as the *Crab Nebula* (fig. 25) emit large quantities of radio waves give us indirect information. It appears that these radio waves can only be produced by relativistic electrons moving in a magnetic field (known as synchroton radiation because this process is observed in a synchroton) and, if relativistic electrons were produced in the explosion, it is likely that relativistic nuclei were also produced: such relativistic nuclei would also radiate energy in a magnetic field but at a much lower rate than electrons.

A supernova is the most violent event of which we have any definite knowledge in our Galaxy, but there are also observations of much more violent explosions in the central regions of some other galaxies (fig. 26). In such an explosion the luminosity of a galaxy may be increased by one or more orders of magnitude. The radio galaxies, mentioned in Chapter 1, appear to be galaxies which have suffered such an explosion. As in the supernova remnants, it appears that relativistic electrons produce the radio waves and it therefore seems possible that cosmic rays are produced in such events. If cosmic rays fill the whole of intergalactic space, they have probably mainly been produced in explosions in central regions of galaxies.

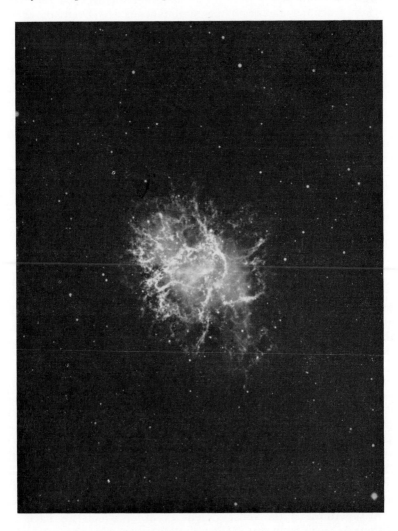

Fig. 25. The Crab Nebula which is the remnant of the supernova of AD 1054. (*Photograph from the Hale Observatories.*)

Life history of cosmic rays

Now let us consider the life history of cosmic rays. Once a cosmic ray leaves its point of origin travelling with almost the speed of light, it would in normal circumstances leave the Galaxy after a time between 10^4 and 10^5 years. If this were true, cosmic rays would escape from the Galaxy almost immediately on an astronomical time-scale and we should have to assume that intergalactic space was filled

Fig. 26. The giant elliptical galaxy in Virgo, M 87, showing a peculiar jet, which is a strong source of radio emission and which may have been produced by an explosion in the central regions of the galaxy. (*Photograph from the Lick Observatory.*)

with cosmic rays. It is not believed that cosmic rays do escape so rapidly from the Galaxy and the reason for this belief is that the Galaxy possesses a magnetic field. In the presence of a magnetic field, the motion of charged particles is modified as described below.

Motion of charged particles in a magnetic field

If we first consider slowly moving particles, so that effects due to the special theory of relativity can be neglected, a particle is acted on by a force $ev_\perp B$, where $v_\perp$ is the component of the particle's velocity perpendicular to the magnetic induction B, e is the particle's charge and the force is perpendicular to both $v_\perp$ and B. This magnetic force causes the particle to move in a circle around a line of magnetic induction, the radius r of the circle being given by

$$mv_\perp^2/r = ev_\perp B,$$

or

$$r = mv_\perp/eB. \tag{4.1}$$

In addition the particle can move freely along the direction of the field lines, so that the total motion is helical as shown in fig. 27.

Fig. 27.   The motion of a charged particle in a magnetic field.   The direction of the magnetic induction B is shown by the solid line.   A charged particle moves on a helical path as shown.

The galactic magnetic field

This simple discussion is complicated when the magnetic field varies in space and time, but the general conclusion is unaltered. The charged particle can travel relatively freely along the direction of the magnetic field, but only a short distance across it. This would scarcely affect the motion of the cosmic rays if a typical magnetic field line passed straight out of the Galaxy. However, to the extent that the structure of the magnetic field in the Galaxy is understood, it appears that the field lines are largely confined to the flattened disk of the Galaxy and that they may circuit the Galaxy several or many times before escaping or that they may even be closed in the Galaxy. This implies that charged particles can be contained in the Galaxy for a time long compared to their free escape time.

The radius of the particle orbit given above is only valid for particles which are moving slowly compared to the velocity of light. It can be generalized in the relativistic case to

$$r = \mathscr{E}/eBc, \tag{4.2}$$

where $\mathscr{E}$ is the total energy of the particle (in joules).   To decide how effectively cosmic rays may be confined to the Galaxy, we need a value for the magnitude of the galactic magnetic field.   A variety of observations, including for example the *Zeeman* splitting* of the 21 cm radio radiation from neutral hydrogen, suggests that the Galaxy contains a magnetic field of strength between 10^{-9} and 10^{-10} tesla.   If we adopt even the largest value suggested for the galactic magnetic field, it is possible to see that the highest energy cosmic rays have orbits with radii comparable with the dimensions of the Galaxy, so that they *cannot possibly* be confined to the Galaxy, but most of the cosmic rays *are* very comfortably confined to the Galaxy.   For instance, in a field of 10^{-9} tesla, a cosmic ray proton of energy 10^{20} eV has an orbit of radius 10^4 parsec, which is comparable with the Earth's distance from the centre of the Galaxy.   Such high energy particles are scarcely affected by the magnetic field.   In contrast, if the energy is 10^{10} eV, the radius is 3×10^{10} m, which is less than the radius of the Earth's orbit around the Sun, and so the lower energy cosmic rays, which are more common, are strongly affected by the magnetic field.

Break-up of cosmic ray nuclei

We can now invert the discussion about the confinement of cosmic rays and ask what would happen to cosmic rays of arbitrary chemical composition if they were confined to the Galaxy indefinitely.   In passing round the Galaxy, the cosmic rays may collide with the inter-stellar gas.   This gas is known, as a result of observations of the type described briefly on page 51, to be primarily hydrogen with a number density of about 10^6 atoms m^{-3}.   The collisions do two things.   They slow down the cosmic rays and they break up the cosmic ray nuclei other than the protons.   The efficiency of break-up is apparently greater than the efficiency of slowing down, so that particles still have cosmic ray energies by the time that they have been broken down into quite light elements.   If the cosmic rays were confined indefinitely, there would be virtually no heavy nuclei left.   The observed distribution of elements shows that this is not the case.   If, as seems likely, the cosmic rays at birth contained a preponderance of heavy nuclei, it is possible to ask how long it would take them to be broken down to give the mass distribution at present observed and, in fact, to see whether the observed mass distribution can be produced by such a process.

The same effect can be obtained by the cosmic rays passing through a dense medium for a short time or through a diffuse medium for a

---

* In a magnetic field, the energy levels of hydrogen (and other atoms), which have been described on page 32 are each split into several closely spaced levels. As a result each spectral line is split into several components.   This is called the Zeeman effect.

much longer time. Results are normally expressed in terms of the mass per unit area to which a cosmic ray has been exposed; that is the mass contained in a cylinder of a square metre in cross-section with the particle's path as axis along the total length of the particle's path. Calculations show that something like the presently observed chemical composition can be obtained if the primary cosmic rays have been exposed to about 40 kg m^{-2} (see Table 10). The light elements Li, Be, B are entirely produced by this break-up process and the requirement that they have their observed abundances relative to their neighbours in the periodic table essentially determines the mass to which the primary cosmic rays have been exposed.

(a) Abundances of the major heavy constituents in the cosmic rays at source, if the observed abundances are produced by passage through 40 kg m^{-2} of matter. The abundances, by number, are relative to carbon normalised to 100.

Element	C	N	O	Ne	Mg	Si	S	Cr	Fe
Abundance	100	12	105	20	23	20	3	8	23

(b) Abundances at the top of the Earth's atmosphere of elements not present in the cosmic rays at source. Abundances produced from the composition given above are compared with observations. Abundances are again normalised to a value 100 for carbon at source.

Element	Li	Be	B	F	P	Cl	K	Sc	Ti	V
Calculated	17	10	27	2·1	0·6	0·7	0·8	0·4	1·8	0·8
Observed	16	11	27	2	0·6	0·5	0·6	0·3	2	1·0

Table 10. In each table the larger numbers are uncertain by about 10 per cent and the smaller ones by up to 50 per cent.

In Table 10 we show an estimate of the abundances of the main heavy constituents of the cosmic rays at source and we also compare the calculated and observed abundances above the Earth's atmosphere of elements not present in the source. Not all of the H and He observed is produced by this break-up process and a proportion of these elements, small compared to their normal stellar abundance, must have been present in the primary cosmic rays. In reality not all cosmic rays we observe have passed through the same amount of matter but the 40 kg m^{-2} is a reasonable average value. There is, of course, no reason why the protons in the cosmic rays should not have passed through much more or much less matter if all of the primary cosmic rays were not produced in similar sources.

If the cosmic rays have been confined to the disk of the Galaxy which contains gas of mass density of order 2×10^{-21} kg m^{-3}, the particle paths must have been about 2×10^{22} m or 2×10^6 light yrs. This suggests that, moving with a speed close to that of light, cosmic rays must be confined to the Galaxy for only a few million years, so that the galactic magnetic field must hold them in for between 10 and 100 times their time of free escape.

Rate of production of cosmic rays

If we assume that cosmic rays stay in the disk of the Galaxy for about 4 million years (twice the figure given above because that is the *average* age of the cosmic ray particles), we can estimate the rate at which cosmic rays must be produced and at which energy is being fed into cosmic rays. Using the energy density of 10^6 eV m^{-3} and the dimensions of the galactic disk given on page 9, the total cosmic ray energy in the disk is of order $1 \cdot 3 \times 10^{43}$ J. If this energy has to be replaced every 4×10^6 yr, the rate of energy input is 10^{34} W. This is almost 10^8 times the total light output of the Sun and it is also of order 10^{-3} of the total light output of the Galaxy. The cosmic rays which escape from our Galaxy and other galaxies produce a small density of cosmic rays in intergalactic space. If most cosmic rays are produced in extremely violent events in a minority of galaxies so that intergalactic space is filled with cosmic rays with a number density similar to that in the Galaxy, the energy input into cosmic rays must be much greater. We shall consider these energy requirements further when we discuss supernovae in Chapter 7.

Significance of cosmic ray abundances

We now return to the connection, if any, which the observed chemical composition of the cosmic rays has with the subject of the origin of the chemical elements. It is believed that the observed abundances tell us more about the life history of the cosmic rays than about their properties at birth. There is, nevertheless, some important information about their birth. The abundances of the heavy elements in the cosmic rays are considerably higher than their terrestrial and solar values. This must have been even more pronounced when they were produced as is indicated by the source composition shown in Table 10. We have mentioned on page 56 three ways in which this overabundance of heavy nuclei may have been produced and it is now thought that the acceleration process is associated with the process actually producing the heavy elements from lighter elements. The suggestion that transuranium elements may be present in the cosmic rays is potentially particularly interesting. All transuranium elements that have yet been produced are unstable (and all are expected to be) and most of them have a very short half-life. If their presence could be confirmed, they would give some direct information about the maximum age of the cosmic rays and would confirm that nucleo-synthesis and production of cosmic rays must be associated. A similar production of transuranium nuclei (but not with relativistic speeds) has been observed in test explosions of nuclear bombs and they are believed to have been produced by the *r*-process of neutron capture which will be described on page 79 of Chapter 5.

Summary of Chapter 4

The chemical composition of the cosmic rays can be studied to a reasonable approximation and it is found that they are richer in heavy elements and in lithium, beryllium and boron than stellar and terrestrial matter. The significance of this composition is not immediately clear. If there were no galactic magnetic field, cosmic rays produced in the Galaxy would escape rapidly into intergalactic space and their observed composition would be similar to their original composition. In fact, they travel in helical paths around the lines of magnetic induction and they remain in the Galaxy long enough to suffer collisions with the interstellar gas which break them up to produce lighter cosmic ray nuclei. If the original chemical composition is rich in heavy nuclei and if cosmic rays are trapped within the Galaxy for about 4×10^6 yr, the abundances resulting from break-up are consistent with observations. The cosmic rays are probably produced in objects such as supernovae, which are also centres of nucleosynthesis of the heavy elements, so that the processes of production of heavy elements and the acceleration of a small fraction to cosmic ray energies are associated. *It should, however, be stressed that many of the steps in the above argument are uncertain and the problem of the origin and life history of cosmic rays is not completely solved.*

theoretical background to the origin of the elements

Introduction

THIS chapter is concerned with identifying the nuclear reactions which have probably played a part in the origin of the chemical elements and with a discussion of the physical conditions in which they are likely to occur. It is not primarily concerned with astrophysics. Astrophysical ideas are introduced occasionally to ensure that too much time is not wasted on a discussion of processes that are irrelevant in astrophysics, but the main discussion of the relevance of particular nuclear reactions to the origin of the elements is reserved for Chapters 6 and 7.

Historical background

In the past there have been several attempts to account for all of the observed elemental and isotopic abundances on the basis of a single process. Such theories did not deny that some modification of initial abundances had been produced by nuclear reactions in stars, but they supposed that such modification had been slight. They were introduced at a time when it was not realised that there were any significant composition differences between stars, so that a single production process seemed very attractive, but even so they were not successful in explaining the general character of the observed abundances. The individual nuclear reactions envisaged in these theories *are* important and it is therefore worthwhile to discuss them briefly.

(a) The αβγ *theory**

This theory was incorporated in the *big-bang* cosmological theory, which will be described in Chapter 6. It was supposed that the Universe was initially very hot and very dense and composed entirely of neutrons. The Universe expanded and cooled and the neutrons gradually decayed into protons. As the protons were formed they captured neutrons to form deuterium, the heavy isotope of hydrogen. It was Gamow's hope that a succession of neutron captures would build up the heavy elements with the abundances that are observed today. For this to happen a large number of neutron captures would have to occur rapidly before all of the neutrons decayed. The

* This is so named from a key paper by Alpher, Bethe and Gamow.

process fails to produce sufficient heavy nuclei because there are no stable nuclei with masses 5 and 8; nuclei of these masses decay very rapidly by emission of a proton or a neutron and it is only rarely possible for an additional neutron to be added to produce a more heavy nucleus. Recent calculations of the early stages of an expanding Universe have shown that essentially only helium can be produced in this manner. The process of rapid neutron capture *is* thought to be important in nucleosynthesis and it will be discussed later in this chapter.

(b) The equilibrium theory

If nuclear reactions are proceeding in a condition close to *thermo-dynamic equilibrium**, a situation is reached in which any nuclear reaction

$$A + B \rightarrow C + D, \tag{5.1}$$

in which the nuclei A, B combine to form the nuclei C, D, is occurring just as rapidly as the reverse reaction

$$C + D \rightarrow A + B. \tag{5.2}$$

When this is true, the abundances of the reacting nuclei depend on the temperature and density of the material alone. At normal stellar temperatures ($10^7 - 10^8$ K), it would take an impossibly long time to establish such an equilibrium, but at temperatures greater than about 4×10^9 K, which are believed to occur in some stars, it might be possible for such an equilibrium to be reached within the star's lifetime. If the temperature at which the equilibrium process is supposed to occur is taken to be about 10^{10} K, some approximation to the observed abundances of the light elements is obtained, but there is again a great lack of heavy nuclei compared with the observations. An equilibrium process *cannot* have been responsible for all of the observed abundances. If the temperature at which equilibrium is reached is less than about 5×10^9 K, the highest abundances occur in the region of iron where the most stable nuclei occur. It is possible that the observed abundance peak in the neighbourhood of iron may have been produced in a near equilibrium process in highly evolved stars whose central temperatures may reach a few thousand million degrees; the equilibrium process is discussed further on page 86.

(c) The polyneutron theory

The two previous theories failed to produce enough heavy elements. In the polyneutron theory it was assumed that initially the matter of the Universe was in the form of a single *primaeval atom* or *polyneutron*. The existing chemical elements were then supposed to have formed by

* For a discussion of the concept of thermodynamic equilibrium see the Appendix on page 162.

fission of the polyneutron. This theory failed to predict the observed preponderance of light elements which is not surprising as the most stable nuclei are ones of intermediate mass such as iron.

None of these theories produces abundances in agreement with observation and it now appears that there are important abundance differences from star to star inside our Galaxy. Some significant alteration of abundances must have occurred during the lifetime of the Galaxy, unless its initial chemical composition was very irregular. In any case a single creation process cannot explain the observations. Because of the overwhelming abundances of the two lightest elements, hydrogen and helium, it is generally assumed that the heavier elements have been produced by build-up from light elements. Although there have been some theories, such as the polyneutron theory above, which suppose that significant fission of heavy elements has occurred, these do not seem very probable, because in any fission process we cannot expect the majority of the material to be converted into nuclei on the low mass side of the peak in the binding energy curve which is discussed in the next section. For the purpose of this book, we will adopt the working hypothesis that a build-up from light elements has occurred.

Energy release from nuclear reactions

Atomic nuclei are composed of protons and neutrons, which together are referred to as nucleons. The total mass of a nucleus is less than the mass of its constituent nucleons. That means that there is a decrease in mass if a compound nucleus is formed from nucleons and, by the Einstein mass-energy relation, $E = mc^2$, this lost mass is released as energy. This energy is known as the *binding energy* of the compound nucleus and it can be calculated when the mass difference between the compound nucleus and its constituent nucleons is known. Thus, if a nucleus is composed of Z protons and N neutrons, its binding energy $Q(Z,N)$ is:

$$Q(Z,N) \equiv (Zm_{\mathrm{p}} + Nm_{\mathrm{n}} - m(Z,N))c^2, \qquad (5.3)$$

where m_{p} is the proton mass, m_{n} the neutron mass and $m(Z,N)$ the mass of the compound nucleus.

A more significant quantity for our present discussion than the total binding energy is the binding energy per nucleon $Q/(Z+N)$. This is proportional to the fractional loss of mass when the compound nucleus is formed. If $Q(Z,N)$ is plotted against $A(=Z+N)$ for a large number of nuclei, the resulting diagram has the character shown in fig. 28. The actual curve is very irregular; in particular for a given value of A there may be several isobars (nuclei which have the same number of nucleons but a different division between protons and neutrons) which have different values of Q. The general property of the curve is that the binding energy per nucleon rises rapidly with the initial increase in nucleon number, there is a broad maximum for A values between 50 and 60 (nuclei in the neighbourhood of iron in the periodic table) and then there is a gradual decline for nuclei with higher values of A. Nuclei containing equal even numbers of both protons and neutrons (excluding ^{8}Be) are found to have higher fractional binding energies than neighbouring nuclei, and particularly strongly bound nuclei are those containing *magic numbers* of protons and neutrons. Although the structure of complicated nuclei is not fully understood, it is observed that nuclei containing 8, 20, 28, 50, 82 or 126 protons or neutrons possess a greater binding energy and stability than their neighbours. They are usually known as *magic number* or *closed shell* nuclei. We shall find that this property is very important when we discuss the origin of the very heavy elements. These nuclear properties are immediately reminiscent of the properties of the observed abundances shown in fig. 24 on page 53. There is an abundance peak in the iron region, there are subsidiary abundance peaks for nuclei with neutron magic numbers or near to these and those nuclei which can be considered to be made up of α particles are more abundant than their neighbours.

From fig. 28 it can be deduced that energy can be released either by the *fusion* of light elements or the *fission* of heavy elements. As mentioned above, it seems likely that only fusion reactions have played an important role in the origin of the elements. Consider the process

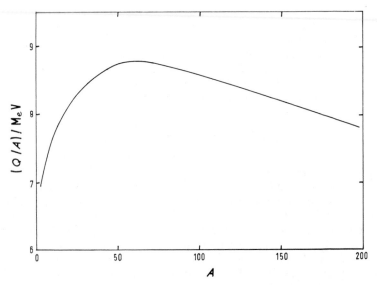

Fig. 28.   The binding energy per nucleon as a function of atomic mass number; Q is measured in MeV.

68

by which light nuclei combine to form heavier nuclei. If two nuclei and their compound all lie to the left of the maximum in fig. 28, the compound nucleus has a larger binding energy per nucleon than the original nuclei and, as the total number of nucleons has not been changed, the nuclear reaction must release energy. Such energy release from nuclear fusion reactions will occur in the conversion of elements such as hydrogen and helium into heavier elements up to and including those in the neighbourhood of iron in the periodic table. Moreover it is known that the energy that stars radiate is largely produced by nuclear fusion reactions in their interiors and that for at least a part of their life history stars' interiors become hotter as they evolve, which leads to nuclear fusion reactions involving successively heavier nuclei. Provided we can explain how the heavy nuclei produced in early generations of stars were expelled into space to be incorporated into later generations of stars, in whose surface layers we observe them, there is at least a possibility that we can account for the production of most elements and isotopes up to iron, cobalt and nickel. The actual nuclear fusion reactions which are thought to occur are discussed later in this chapter and the gradual change in composition of the galactic gas caused by the reactions is discussed in Chapter 7.

The production of very heavy elements

If it is accepted that the originally created matter was entirely light elements, there is no such simple explanation of the occurrence of elements heavier than nickel. Their production requires an input of energy rather than yielding energy, given that many particle nuclear reactions involving large numbers of light elements are extremely unlikely*. In addition even with an input of energy, it becomes extremely difficult to produce such heavy nuclei by reactions involving two charged nuclei. The repulsive electromagnetic interaction between two charged nuclei tends to prevent them from approaching close enough for the nuclear interaction to cause a nuclear reaction. This repulsive interaction increases as the product of the charges of the two nuclei and, for heavy highly charged nuclei, reactions can only occur for relative velocities approaching the speed of light.

Neutron capture reactions

It is generally agreed that the nuclei heavier than nickel, and possibly some lighter nuclei, have been produced by a process of successive *neutron capture*, which has some similarity with the $\alpha\beta\gamma$ process mentioned earlier. Provided a supply of neutrons is available, these neutron capture reactions do not share the difficulties of the reactions

* For example, the isotope ^{56}Ni will not be produced by simultaneous interaction of 14 α particles.

between two charged particles, as neutrons have no electric charge and are not repelled by the other nucleus. We can first ask ourselves what happens if material is irradiated with fluxes of neutrons of arbitrary energy for a variety of periods of time. If for some choice of flux, energy and time, the elemental and isotopic abundances produced are similar to those which occur naturally, we must then look for some place in the Universe where such a source of neutrons might exist. In so doing we must remember that neutrons are unstable particles with a half life of a little over 10 minutes, so that they must be used almost as soon as they are produced.

The first thing to be said about any neutron capture process is that very frequently the nucleus produced by neutron capture will be unstable to β-decay. Thus the reaction

$$(Z,A) + n = (Z,A+1) \tag{5.4}$$

will frequently be followed by

$$(Z,A+1) \rightarrow (Z+1,\ A+1) + e^- + \bar{\nu}, \tag{5.5}$$

where $\bar{\nu}$ denotes an anti-neutrino. A specific example is the reaction

$$^{58}\text{Fe} + n \rightarrow {}^{59}\text{Fe}, \tag{5.6}$$

which is followed by

$$^{59}\text{Fe} \rightarrow {}^{59}\text{Co} + e^- + \bar{\nu}. \tag{5.7}$$

The progress of any process of neutron capture will depend on whether a further neutron is captured by the nucleus $(Z, A+1)$ before it is able to β-decay or whether the nucleus decays before it captures the next neutron. Obviously it will be possible to arrange a flux of neutrons which is such that some unstable product nuclei decay before further neutron capture while others do not, but the variations in β-decay rates and neutron capture rates are less than the possible variation in neutron fluxes and generally capture is supposed to occur through two distinct processes:

(i) *The s-process.* This is *slow neutron capture* in which any unstable nucleus always decays before capturing a further neutron.

(ii) *The r-process.* This is *rapid neutron capture* in which several or many neutrons are captured before the product nucleus becomes so unstable that no further neutron can be captured and decay must occur.

It might be thought that there would always be a mixture of the two processes, but if it is nucleosynthesis inside stars that is being discussed, there is a reason why there is likely to be a reasonably sharp division between them. During normal stellar evolution there are three characteristic times (*timescales*) over which the properties of a star may change significantly (see *The Stars* for more detail). They are:

(i) The *dynamical timescale*

$$t_d \approx (r_s{}^3/GM_s)^{\frac{1}{2}}, \qquad (5.8)$$

where r_s and M_s are the radius and mass of the star and G is the Newtonian gravitational constant ($6 \cdot 67 \times 10^{-11} \, \mathrm{N \, m^2 \, kg^{-2}}$).

(ii) The *thermal timescale*

$$t_{th} \approx GM_s{}^2/L_s r_s, \qquad (5.9)$$

where L_s is the luminosity of the star.

(iii) The *nuclear timescale*

$$t_n \approx \mathscr{E}_n/L_s, \qquad (5.10)$$

where $\mathscr{E}_n$ is the total energy which can be released by whatever nuclear reaction is supplying the star's energy.

The dynamical timescale is the time that would be needed for readjustment, if the forces involved in the equilibrium of a star were seriously out of balance, and it is specifically the timescale of explosive events in stars such as supernovae. Most stages of stellar evolution are, in contrast, very slow, proceeding either on a nuclear timescale or, on occasions when a nuclear fuel has been exhausted, on a thermal timescale. In most stars

$$t_n \gg t_{th} \gg t_d \qquad (5.11)$$

and a typical dynamical timescale is $10^3 - 10^4$ s, while thermal timescales usually range upward from 10^4 yr. If neutron capture reactions are initiated by neutrons released by nuclear reactions during a normal evolutionary phase of a star proceeding on a thermal or nuclear timescale, β-decay of unstable nuclei will certainly be able to occur between successive captures of neutrons and an *s*-process will necessarily result. If neutron capture reactions are caused by neutrons released during a stellar explosion, it will lead to an *r*-process. The great difference between t_{th} and t_d and the improbability that neutrons will be supplied on a timescale intermediate between them makes these processes essentially distinct.

It is possible to indicate which isotopes of elements heavier than iron should have been produced by each of these processes. Some isotopes can only be produced by one of the two processes, while others can be made by either and there are apparently some isotopes which can be made by neither. Further mention of this will be made later. The manner in which isotopes can be attributed to the two

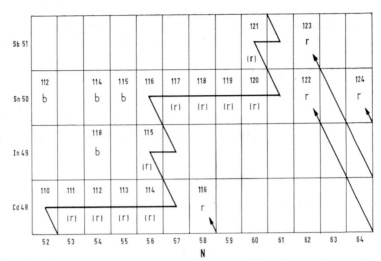

Fig. 29. Neutron capture processes in the neighbourhood of tin. In this chart the number of neutrons increases to the right and the number of protons upwards. The stable isotopes are marked *r* if they are produced by the *r*-process, (*r*) if they can be produced by both *r*- and *s*-processes, *b* if they are by-passed and they are unmarked if they are only produced by the *s*-process. The solid line shows the *s*-process path. The arrows show how β-decays lead to the *r*-process isotopes.

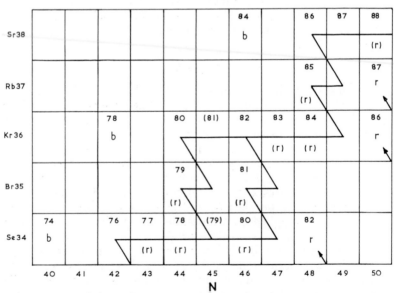

Fig. 30. Neutron capture processes in the neighbourhood of krypton. The notation is as in fig. 29. Note that the *s*-process is somewhat unusual. [79]Se is unstable but long-lived enough to capture an additional neutron sometimes and the *s*-process path splits at this point to join again at [82]Kr.

72

processes in two regions of the periodic table are shown in figs. 29 and 30. Roughly speaking it can be said that neutron rich isotopes will be produced by the r-process, whereas those isotopes produced by the s-process will have relatively larger numbers of protons. Those that can be produced by neither neutron capture chain are the isotopes with the largest number of protons and these are generally known as *bypassed nuclei*. The bypassed nuclei are also shown in figs. 29 and 30. Originally it was suggested that they might be produced by proton capture (p-process), but at present there is no completely satisfactory theory of their production. They are typically isotopes of very low relative abundance and this gives support to the view that neutron capture is the main process involved in the production of the heavy elements and that only a rare or inefficient process is required to produce the remaining isotopes.

The isotopes of tin

The element tin is of particular interest as it has more stable isotopes than any other element; this is related to the fact that it contains a magic number, 50, of protons. The relative abundances of the stable isotopes of tin in terrestrial and meteoritic matter are shown in Table 11. The three lightest isotopes cannot be made by the r- or s-process but they only contribute 2 per cent of the total abundance. The isotope ^{116}Sn can only be made by the s-process and ^{122}Sn and ^{124}Sn can only be made by the r-process. The four remaining isotopes have contributions from both the s- and r-process. Some further comments about the abundances of the isotopes of tin will be made below.

Nuclear mass number	112	114	115	116	117	118	119	120	122	124
Abundance	1·02	0·69	0·38	14·3	7·6	24·1	8·5	32·5	4·8	6·1
Process	b	b	b	s	s,r	s,r	s,r	s,r	r	r

Table 11. The relative abundances of the stable isotopes of tin. The abundances of the ten stable isotopes are shown as a percentage of the total. It is also shown whether the isotopes are bypassed (b) or produced by the s-process (s) or the r-process (r).

Neutron capture cross-sections

The rate of nuclear reactions is usually expressed in terms of a quantity known as the *reaction cross-section*. Suppose we imagine a beam of incident particles to fall on a target (fig. 31). If the target particle behaved like a billiard ball, it would collide with all of those particles in the incident beam which lie in a cylinder of cross-section πa^2, where a is the radius of the target particle. In describing a process involving real particles, it is convenient to introduce an effective cross-section of the target particles, which indicates what area of the incident beam is affected. If the velocity of the incident particles is v, the cross-section for collisions is σ and the numbers of incident and target particles per cubic metre are n_I and n_T, the number of collisions in unit volume in unit time is

$$\text{number of collisions} = n_I n_T \sigma v. \tag{5.12}$$

We shall need to use expressions like (5.12) in what follows. Usually, except in a laboratory experiment, there is no clear distinction between incident and target particles but this does not matter provided that the *relative* velocity of the two particles is inserted in (5.12). In addition the velocities of particles are not usually all the same and σ depends on v, so that (5.12) has to be generalized by integrating over all values of the relative velocity of the particles. It must be noted that capture is not the only possible outcome of interaction between a neutron and a heavy nucleus; it may simply have its direction of motion altered and a *scattering cross-section* measures the probability of this process.

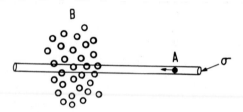

Fig. 31. Reaction cross-section. A particle A is projected towards a target B. The number of reactions is the same as if the particle reacted with all target particles in a cylinder of cross-sectional area σ centred on itself.

The neutron capture cross-sections have the property that to a high degree of accuracy

$$\sigma \propto 1/v. \tag{5.13}$$

The relation is approximately true for the whole range of velocities which is believed to be important in the r- and s-processes and the approximation is improved when the average over relevant velocities is performed. This means that in discussing neutron capture we will take σv to be a constant. If we can obtain a typical value of σv, then we can discover what density of neutrons will be required to produce significant nuclear transmutation. Cross-sections are not usually expressed in SI units. Nuclear and neutron reaction cross-sections are usually expressed in terms of a unit called a barn, where

$$1 \text{ barn} = 10^{-28} \text{ m}^2. \tag{5.14}$$

Most of the cross-sections are a fraction of a barn. In the case of neutron capture reactions, for a typical nucleus of large mass, a characteristic value of σv is

$$\sigma v \approx 3 \times 10^{-23} \mathrm{m^3 s^{-1}}. \tag{5.15}$$

If there are n_n neutrons per cubic metre, a nucleus will capture one neutron in a time τ_n which is approximately given by

$$\tau_\mathrm{n} = 1/n_\mathrm{n}\sigma v = 3 \times 10^{22}/n_\mathrm{n} \ \mathrm{s}$$
$$\approx 10^{15}/n_\mathrm{n} \ \mathrm{yr}. \tag{5.16}$$

We have previously mentioned that we expect the s-process to occur in times of order 10^4 yr or more. For significant neutron capture to occur in such a time, we need upwards of 10^{11} neutrons per cubic metre. In typical stellar densities which range upwards from 10^3 kg m^{-3} (6×10^{29} nucleons m^{-3}) such a number of neutrons would be a very minor constituent of the matter. In the r-process we require neutron capture to occur in a fraction of a stellar dynamical time so that many neutrons can be captured by a target nucleus during the course of the explosion. Capture in one second requires 3×10^{22} neutrons m^{-3} and some calculations of the r-process have required up to 10^{30} neutrons m^{-3} to produce observed abundances. Such neutron densities can only be obtained if a significant fraction of the matter present is converted into neutrons and in a very short time.

Details of the s-process

We now consider the details of the s-process and we make the approximation (5.11) that σv is constant. The first point about the s-process is that, starting with a given initial nucleus, there will be just one nucleus which can be produced by the s-process for each value of A, the number of nucleons in the nucleus; the nucleus produced by neutron capture, which increases A by unity, is either itself stable or it decays to a nucleus that is stable. Suppose that an s-process is under way with n_n neutrons per cubic metre. Then the rate of production of nuclei of nucleon number A can be described by the equation

$$\frac{dn_A}{dt} = n_\mathrm{n}(\sigma_{A-1}v)n_{A-1} - n_\mathrm{n}(\sigma_A v)n_A, \tag{5.17}$$

where the first term on the right-hand side represents the production of nuclei of type A from those of type $A-1$, whilst the second term represents the production of those of type $A+1$ from type A. Since σv is assumed constant, we may write

$$\left. \begin{array}{l} \sigma_{A-1}v = \bar\sigma_{A-1}\bar v, \\[2mm] \sigma_A v = \bar\sigma_A \bar v, \end{array} \right\} \tag{5.18}$$

75

where $\bar{v}$ is the mean velocity of the neutrons which will be related to the temperature, T, of the matter including the neutrons and $\bar{\sigma}_{A-1}$, $\bar{\sigma}_A$ are mean values of the cross-sections. By introducing a variable τ defined so that

$$d\tau = n_n \bar{v} dt, \qquad (5.19)$$

eqn. (5.17) can be written

$$\frac{dn_A}{d\tau} = \bar{\sigma}_{A-1} n_{A-1} - \bar{\sigma}_A n_A. \qquad (5.20)$$

If n_n and $\bar{v}$ can be regarded as approximately constant while the neutron capture is proceeding, τ is simply a multiple of t; otherwise eqn. (5.19) must be integrated to obtain τ when the time variation of n_n and $\bar{v}$ is known.

If an s-process is proceeding steadily there is a probability that, in the middle of the reaction chain, the net rate of change of abundances will be small compared to the rate at which individual nuclei progress along the reaction chain. In that case the two terms on the right-hand side of (5.20) would be individually much larger than the left-hand side and we would have

$$\bar{\sigma}_{A-1} n_{A-1} \approx \bar{\sigma}_A n_A. \qquad (5.21)$$

Of course, this approximate equation cannot possibly be true for nuclei near the beginning and end of the reaction chain and it is also unlikely to be true for the magic number nuclei, which were mentioned earlier. These nuclei are particularly stable and have very low cross-sections for capturing another neutron. Because of this low cross-section, it is less likely that the two terms on the right-hand side of an equation like (5.20) can both be much larger than the left-hand side.

Termination of the s-process

The end of the reaction chain has just been mentioned and there is, in fact, a genuine end to the s-process. All nuclei more massive than ^{209}Bi are unstable, many of them to α emission rather than to β-decay and ^{209}Bi is the heaviest isotope that can be produced by the s-process. There is, of course, no reason why ^{209}Bi should not capture another neutron but, if it does, it decays by emission of an α particle to ^{206}Pb. This means that the equation for the rate of change of the abundance of ^{206}Pb is not (5.20) but is instead

$$\frac{dn_{206}}{d\tau} = \bar{\sigma}_{205} n_{205} - \bar{\sigma}_{206} n_{206} + \bar{\sigma}_{209} n_{209}. \qquad (5.22)$$

This then gives a closed set of equations from which s-process abundances can be calculated.

Calculations of s-process

Because the abundance of iron is very much greater than that of elements a little less massive and of all heavier elements, calculations of the s-process normally proceed as follows. A flux of neutrons is assumed to irradiate pure iron (and usually pure ^{56}Fe, the most abundant isotope of iron in terrestrial and meteoritic samples) and the rate of change of the abundances of the heavier s-process isotopes is studied. The results are given in terms of τ which is called the *neutron exposure* rather than t. The calculated results for several different values of neutron exposure are shown in fig. 32. Several properties are apparent in these diagrams. There is the gradual disappearance of iron nuclei and those a little heavier and the gradual build-up to the really heavy nuclei as τ increases. As predicted by our simple argument leading to eqn. (5.21), there are regions in which $\bar{\sigma}n$ is approximately constant but these regions tend to be separated by small ranges of values of A where $\bar{\sigma}n$ changes very rapidly. The rapid changes occur close to the magic number nuclei.

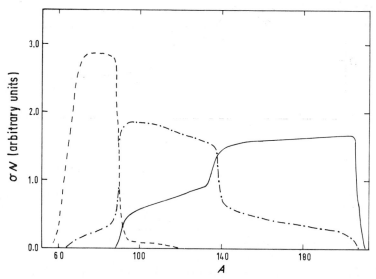

Fig. 32. The s-process abundances produced by three different neutron exposures on iron. The three curves from left to right correspond to increasing neutron exposure, with the resulting build-up of heavy nuclei.

Evidence for occurrence of s-process

No single one of these curves reproduces the solar system abundances of those isotopes which are believed to have been produced by the s-process. The observed abundances, which have already been shown in fig. 24 and which are derived from a study of the Earth and meteorites, are plotted in fig. 33. It can be seen that there are regions in which $\bar{\sigma}n$ is approximately constant but that over the entire mass range $\bar{\sigma}n$ varies by several orders of magnitude. Calculations have shown that, if the heavy elements in the solar system have been processed in several different places by s-processes with different values of τ and then mixed up before the formation of the Sun and planets, it is possible to reproduce the observed abundances. The solid curve in fig. 33 shows the result of such a calculation.

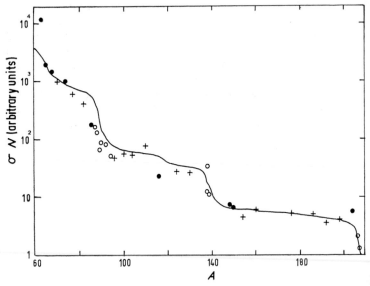

Fig. 33. The solar system s-process abundances. The observed values of σN are shown. The filled circles are isotopes produced by the s-process only, with a reliable cross-section. The open circles also have a reliable σ but are partly produced by the r-process and a contribution due to this has been subtracted. The crosses are isotopes produced only by the s-process but with estimated cross-sections. The solid curve is a theoretical result from a superposition of many s-processes with different values of τ.

It is generally agreed that the idea of the s-process is confirmed by the abundances of the isotopes of several elements such as tin and samarium. As mentioned above, tin has 10 stable isotopes of which five are thought to be wholly or partly formed by the s-process. $\bar{\sigma}n$ is shown in Table 12 for the stable isotopes of tin other than the bypassed isotopes. The values of $\bar{\sigma}n$ for those isotopes produced purely by the r-process are much less than for the other isotopes. $\bar{\sigma}n$ is approximately the same for three of the isotopes which are partly or wholly produced by the s-process; the higher values of $\bar{\sigma}n$ for the other two isotopes probably indicate that the r-process also contributes significantly to their abundance. This can be tested once calculations have been made of the *r*-process. A second observation in favour of the *s*-process is that spectral lines of *technetium* have been seen in stars. Technetium is an element with *no stable isotopes* and the longest half-life is 2.6×10^6 yr, which is much shorter than the lifetime of the stars. The technetium *must* have been produced in the star in which it is observed. It has a long enough half-life to be regarded as stable in the calculation of the *s*-process by which its production is predicted.

Nuclear mass number	116	117	118	119	120	122	124
$\bar{\sigma}$	104	418	65	257	41	~45	~36
n	14·3	7·6	24·1	8·5	32·5	4·8	6·1
$\bar{\sigma}n$	1487	3177	1586	2184	1333	~216	~220

Table 12. Product of cross-section for neutron capture and isotopic abundance for isotopes of tin. $\bar{\sigma}n$ is shown with $\bar{\sigma}$ measured in 10^{-3} barns (10^{-31} m^2) and n the percentage abundance shown in Table 11. No results are shown for the three lightest isotopes whose abundance is so small that it has not proved possible to measure their neutron capture cross-section.

Details of the r-process

The *r*-process is in detail rather more complicated than the *s*-process but we give a discussion which concentrates on the most important features. Before the *r*-process is described, we must discuss nuclear binding energy a little further. If we add an additional proton or neutron to a stable nucleus, the total binding energy of the new nucleus will be greater than that of the old, although the binding energy per nucleon will probably be less. As a result, the new nucleus may be unstable, but it will exist for a finite time before it decays. If we add more and more neutrons (say) to the nucleus, so rapidly that β-decay cannot occur, the neutron excess will eventually become so large that the addition of one further neutron would lead to a nucleus of decreased total binding energy. At this stage no further neutrons can be added to the nucleus and the instability of the nucleus is complete. Unfortunately the binding energy can only be measured for stable or slightly unstable nuclei and there is no generally accepted theoretical model which calculates the binding energy of very unstable nuclei. This means that it is not possible to give a precise value for the total number of neutrons which can be captured by a nucleus before no more can be added. The best estimates of the masses and hence binding energies of heavy unstable nuclei involve the use of the von Weizsäcker semi-empirical mass law. This is a formula which gives the masses of stable nuclei to a reasonable approximation in terms of the numbers of neutrons and protons they contain and of other properties such as their nearness to magic numbers. This formula, which fits the observed nuclei, can be extrapolated with moderate confidence to the very unstable nuclei.

In the simplest version of the r-process a nucleus undergoes rapid neutron capture until it can capture no further neutrons. At this stage there is a *waiting point* until the nucleus undergoes β-decay, thus increasing Z by one and decreasing N by one. The new nucleus can probably now capture additional neutrons until it reaches another waiting point. Assuming that neutron capture is extremely rapid in the r-process chain, we must consider one nucleus at each value of Z and the equations governing the chain have the form

$$\frac{dn_Z}{dt} = \lambda_{Z-1} n_{Z-1} - \lambda_Z n_Z, \qquad (5.23)$$

where λ_Z is the β-decay rate at the waiting point for charge Z. This is the reciprocal of the time it would take n_Z to decrease by a factor e if there were no production of new Z nuclei. Thus, if $n_{Z-1} = 0$, n_Z decays exponentially according to the formula

$$n_Z = n_{Z0} \exp(-\lambda_Z t). \qquad (5.24)$$

The eqns. (5.23) are mathematically essentially the same as eqns. (5.20) but, in the r-process, abundances will be greatest where β-decay rates are lowest. Note also that, whereas the nuclei that enter into eqn. (5.20) are the final nuclei that are observed in nature, the nuclei whose abundances are determined by (5.23) are unstable and, once neutron exposure has ceased, they will decay until a stable nucleus is reached. There is no obvious end-point to the r-process which does not stop when there are no longer any stable nuclei. Thus, whereas the s-process can produce no nuclei more massive than ^{209}Bi, the r-process can produce radioactive nuclei up to and beyond uranium. We cannot know what is the most massive nucleus which has been produced in an r-process but we can feel certain that the solar system material originally contained elements beyond uranium.

There are complications to the simple picture of the r-process given above which cannot be discussed here, but the following points can be made. Nuclei with a magic number of neutrons (50, 82 or 126) are more stable than their neighbours and waiting points are likely to occur at magic numbers. In addition magic number nuclei have rather long β-decay times and they are therefore likely to have large abundances. In the solar system abundances shown in fig. 24 on page 53, there are pairs of abundance peaks near neutron magic numbers. The one with the larger A (and hence Z) is believed to be due to the s-process and the one with the smaller A to the r-process. Because abundances in the r-process are determined by values of β-decay rates rather than neutron capture cross-sections, we do not expect any regularity in a plot of σn against A for nuclei produced by this process; the observations shown in fig. 34 confirm this.   Actual observations of the results of the r-process shortly after it has occurred have been made in nuclear bomb explosions.   In such an explosion there is a high flux of neutrons and isotopes, which are predicted to be made in the r-process, have been identified in the debris.   It is, of course, an unusual r-process as the bomb contains uranium which acts as a target particle for the neutrons.

Calculations can be made of the r-process in the same way as the calculations described for the s-process.   Results of such a calculation are shown in fig. 35.   The very high magic number peaks are apparent but abundances do not vary by a very large factor between these

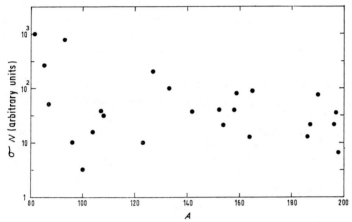

Fig. 34.   The solar system r-process abundances.   The values of σN for the r-process isotopes show no regularity, a result which is expected.

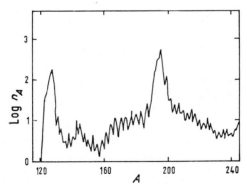

Fig. 35.   A calculation of the r-process showing large peaks near neutron magic numbers and the more gradual change of abundances between magic numbers.

81

peaks. It seems clear that it is possible to account for observed abundances by a mixture of such processes but the biggest uncertainty at present is where in nature the process has occurred. This will be discussed briefly in Chapter 7.

The p-process?

We mentioned earlier that there are some isotopes of the heavy elements which cannot be produced by either neutron capture process. These are the most proton rich isotopes such as ^{112}Sn, ^{114}Sn and ^{115}Sn. It has been suggested that they have been formed by a process of *rapid proton capture* and it is possible that this is the true explanation. The theory is not really satisfactory as there are no features of the observations which point definitely to the necessity of proton capture and, even more than in the case of neutron capture, there is no obvious site in which such reactions have occurred.

Thermonuclear reactions and the production of the elements up to iron

We have now discussed at some considerable length the way in which we believe that the very heavy elements have been produced. These represent only a minute fraction of the total amount of matter of which we have knowledge, and to date we have said virtually nothing about the production of the lighter elements, other than that they can be produced by nuclear fusion reactions releasing energy. It is usually supposed that these reactions take place inside stars, but the site of the nucleosynthesis is not vital for the general discussion.

Nuclei interact through two strong forces. Because they bear positive charges, they repel one another and describe hyperbolic orbits relative to their centre of mass. They also interact through the *strong nuclear interaction* which is basically attractive but which has an extremely short range. It is the strong nuclear interaction which leads to nuclear fusion reactions and which holds nuclei together. If two nuclei are to get close enough together to fuse, despite the repulsive electrostatic force, they must be moving very rapidly. If the velocities of the nuclei are those which they possess because they form a gas of temperature T, high velocities are equivalent to high temperature. This leads to the concept of *thermonuclear reactions**. The rate of thermonuclear reactions depends most sensitively on temperature but it also depends on the density, ρ, of matter. For two particle nuclear reactions in which two nuclei react to form one or more product nuclei, the number of reactions per unit volume is proportional to ρ^2.

If we imagine that we start with a gas of given chemical composition at a relatively low temperature, we can ask what nuclear reactions will

* This is an abbreviated version of a discussion given in *The Stars*.

occur if the temperature of the gas is increased. What happens will depend slightly on the density of the gas but this dependence can be neglected to a first approximation. As the material in the Universe appears to be almost entirely hydrogen and helium, it is natural to assume that they are the principal components in the gas we are considering. As the temperature is increased, the first important series of nuclear reactions to occur is the conversion of hydrogen into helium which can be summarized by

$$4^1\text{H} \rightarrow {}^4\text{He} + 2e^+ + 2\nu; \qquad (5.25)$$

four protons are converted into an α particle, two positrons and two neutrinos.

Hydrogen burning reactions

This process is usually referred to as *hydrogen burning*. There are two different reaction chains, the PP chain in which hydrogen is converted directly into helium and the CN cycle in which nuclei of carbon and nitrogen are used as catalysts. Details of the PP chain are shown in expressions (5.26)–(5.28) and of the CN cycle in expression (5.29). The proton-proton chain divides into three main branches which are called the PP I, PP II and PP III chains; other possible reactions between light nuclei have been shown to be much less probable than those that are shown*. In the expressions for the PP chains, the notation d is used for a deuteron and γ for a photon.

$$
\begin{array}{lll}
 & \textbf{PP I Chain} & \\
1 & \text{p(p, } e^+ + \nu)\text{d} & \\
2 & \text{d(p, } \gamma)^3\text{He} & \left.\vphantom{\begin{array}{c}1\\2\\3\end{array}}\right\} \quad (5.26) \\
3 & {}^3\text{He}({}^3\text{He, p} + \text{p})^4\text{He} &
\end{array}
$$

$$
\begin{array}{lll}
 & \textbf{PP II Chian} & \\
 & \text{This starts with reactions 1 and 2} & \\
\text{Then} \quad 3' & {}^3\text{He}({}^4\text{He, } \gamma)^7\text{Be} & \\
4' & {}^7\text{Be}(e^-, \nu)^7\text{Li} & \left.\vphantom{\begin{array}{c}1\\2\\3\end{array}}\right\} \quad (5.27) \\
5' & {}^7\text{Li}(\text{p}, \alpha)^4\text{He} &
\end{array}
$$

* In the hydrogen burning reaction chains an abbreviated notation has been used for reactions such as those which we have previously written $A + B \rightarrow C + D$. This can be written $A(B,C)D$ where the initially present particles are written to the left of the comma and the final particles to the right.

The CN cycle cannot occur if there is no carbon or nitrogen present in the material but the PPI chain will work in material which is initially pure hydrogen. At densities which typically occur in stars, hydrogen burning occurs at temperatures between 10^7 and 2×10^7 K. At these temperatures and stellar densities, it takes between 10^6 and 10^{10} yrs for a substantial amount of hydrogen to be burnt, so that it is a very slow process. If, in contrast, hydrogen could be brought instantaneously to a temperature of 10^9 K, it would burn *explosively* as the nuclear reaction rates depend very strongly on temperature.

Helium burning

If the temperature of the material rises above 2×10^7 K after all of the hydrogen has been converted into helium, no further significant reactions will occur until the temperature is between 10^8 and 2×10^8 K. At this stage helium burning will occur through the reaction

$$3\ {}^{4}\text{He} \rightarrow {}^{12}\text{C} + \gamma \qquad (5.30)$$

This is a reaction involving three helium nuclei because the reaction product of two helium nuclei, ^{8}Be, is highly unstable. As the carbon is produced, while there is still some helium remaining, there is the possibility of the reaction

$$ {}^{12}\text{C} + {}^{4}\text{He} \rightarrow {}^{16}\text{O} + \gamma, \qquad (5.31)$$

84

leading to the production of oxygen. It used to be thought that the reaction (5.31) would be followed by

$$^{16}\text{O} + {}^4\text{He} \rightarrow {}^{20}\text{Ne} + \gamma, \tag{5.32}$$

but it now appears that the rate of reaction (5.32) is too low for it to be important, all of the helium having disappeared in the production of carbon and oxygen before neon is formed. Thus the net result of helium burning is a mixture of carbon and oxygen, the relative amounts of the two depending on the temperature of helium burning. In principle the production of carbon and oxygen can be calculated as a function of reaction density and temperature. In practice it is not that easy as the reaction cross-sections are not known precisely as it is difficult to measure them at the energies that are important in stars. This means that there is some uncertainty in the results of helium burning, which is important when the next set of nuclear reactions is considered.

Carbon and oxygen burning

Because the charges on carbon and oxygen nuclei are quite high (6 and 8), there must be another substantial increase in temperature before further reactions occur. At temperatures around 5×10^8 K to 10^9 K, carbon and oxygen begin to burn through reactions like

$$\left. \begin{aligned} ^{12}\text{C} + {}^{12}\text{C} &\rightarrow {}^{24}\text{Mg} + \gamma, \\ &\rightarrow {}^{23}\text{Na} + \text{p}, \\ &\rightarrow {}^{20}\text{Ne} + \alpha, \end{aligned} \right\} \tag{5.33}$$

and

$$\left. \begin{aligned} ^{16}\text{O} + {}^{16}\text{O} &\rightarrow {}^{32}\text{S} \ \ + \gamma, \\ &\rightarrow {}^{31}\text{P} \ \ + \text{p}, \\ &\rightarrow {}^{31}\text{S} \ \ + \text{n}, \\ &\rightarrow {}^{28}\text{Si} \ \ + \alpha. \end{aligned} \right\} \tag{5.34}$$

Carbon burning starts at about 5×10^8 K and oxygen burning at 10^9 K. An important property of reactions (5.33) and (5.34) is that there is a large number of possible reactions, which means that a discussion of the results of the reactions is more difficult than in the case of hydrogen and helium burning. This property of many competing reactions continues until the heavier nuclei up to iron and nickel have been produced. The protons, neutrons and α particles released in the reactions (5.33) and (5.34) rapidly undergo further reactions and typically the main product of carbon and oxygen burning is ^{28}Si, which is a particularly strongly bound nucleus.

Silicon burning

Once nuclei in the region of silicon have been produced, the nuclear charges are so high that it is unlikely that reactions like

$$^{28}\text{Si} + {}^{28}\text{Si} \rightarrow {}^{56}\text{Ni} + \gamma \qquad (5.35)$$

will occur. The relatively heavier elements must be produced in a slightly different manner. As the temperature rises to 2×10^9 K and above, the gamma rays which are present in the thermal radiation (at this temperature a typical photon has an energy of 2×10^5 eV and some are much more energetic) become energetic enough to remove α particles and protons from the nuclei of elements like magnesium, silicon and sulphur which are present. These α particles and protons then combine with other nuclei to produce heavier nuclei. Thus symbolically we can replace reaction (5.35) by

$$\left.\begin{array}{l} {}^{28}\text{Si} + \gamma\text{'s} \quad \rightarrow 7\,{}^4\text{He}, \\ {}^{28}\text{Si} + 7\,{}^4\text{He} \rightarrow {}^{56}\text{Ni}. \end{array}\right\} \qquad (5.36)$$

This process whereby some nuclei are broken down and others are built up is given the name of *silicon burning*, although other nuclei than silicon are involved in the process. Since the binding energy per nucleon increases until the region of iron and nickel is reached, there is a net energy release in reactions like (5.36).

Nuclear statistical equilibrium

There are many competing nuclear reactions involved in silicon burning and not all of the reaction cross-sections are well known. This means that a precise calculation of the results of silicon burning is difficult; fig. 36 shows what reactions have been studied in calculations, the *reaction network* for silicon burning. There is, however, a possible simplification. As the temperature rises still further to the neighbourhood of 4×10^9 K, the nuclear reactions become extremely frequent and a situation approaching *nuclear statistical equilibrium* is reached. Usually we can assume that the chemical composition of material can be specified and that its gradual change due to nuclear reactions can be calculated. Eventually, if the material were left for a long time at the same temperature, the chemical composition would reach a statistically steady state*. Although further nuclear reactions would occur, a reaction such as

$$A + B \rightleftharpoons C + D \qquad (5.37)$$

involving four nuclei A, B, C, D would occur equally frequently in each direction. When equilibrium was reached, the nuclear abundances would then depend on ρ and T alone. At low

* See the Appendix on page 162 for a discussion of thermodynamic equilibrium.

temperatures the approach to such an equilibrium would take vastly longer than the believed age of the Universe but at the temperatures which we are discussing here an approach to equilibrium can occur in a year or even much less. On page 66 we have already mentioned that such an *equilibrium process* acting at an even higher temperature fails to account simultaneously for the abundances of both light and heavy nuclei.

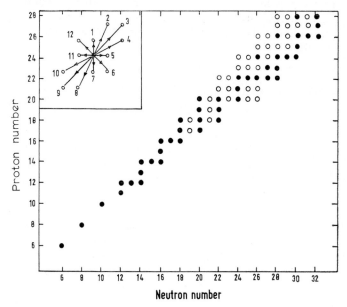

Fig. 36. The reaction network for silicon burning. A recent calculation of silicon burning has included all of the nuclei shown in this grid; the filled circles are stable nuclei and the open circles are β-unstable nuclei. Twelve types of reaction have been included when they connect nuclei which are both in the grid and they are indicated in the inset. The processes are 1 (p,γ), 2 (α,n), 3 (α,γ), 4 (α,p), 5 (n,γ), 6 (n,p), 7 (γ,p), 8 (n,α), 9 (γ,α), 10 (p,α), 11 (γ,n) 12 (p,n), where in each reaction the particle to the left of the comma is added to the target nucleus and that to the right is ejected.

This discussion is somewhat oversimplified. Ordinary nuclear reactions occur very rapidly at high temperatures because the high speed of the particles increases the probability that they will get close enough for nuclear reactions to occur at all. In contrast the β-decay of an unstable nucleus is unaffected by the temperature and density of the matter unless they are even more extreme than those that we are considering. If a true statistical equilibrium is to be reached from an arbitrary initial composition, some β-decay reactions may be needed to change neutrons into protons or conversely. If the β-decays are not sufficiently rapid, the equilibrium which is approached may not be the complete equilibrium but may be only a partial one in which the total numbers of both protons and neutrons present is the same as when the process started.

A second reason for lack of equilibrium is that true thermodynamic equilibrium can only be reached in a self-contained system which is exchanging neither matter nor energy with its surroundings. At high temperatures it is known that in addition to nuclear reactions there occur reactions leading to the production of neutrinos and anti-neutrinos. These reactions include the creation of *electron–positron pairs* from photons

$$\gamma + \gamma \to e^- + e^+, \tag{5.38}$$

which can be followed (rarely) by the reaction

$$e^- + e^+ \to \nu + \bar{\nu} \tag{5.39}$$

producing a *neutrino–antineutrino pair*. If we are considering nuclear reactions inside a star these neutrinos escape ensuring that the star is not a self-contained system and that true equilibrium cannot be reached.

Abundances produced by the equilibrium process

It is possible to calculate what nuclei would be present if either complete equilibrium or the incomplete equilibrium mentioned above were reached. Usually equilibrium favours the most strongly bound nuclei in the neighbourhood of the iron peak and the presence of the strong overabundance of iron and its neighbours in the observed abundances described in Chapters 2 and 3 has led to the belief that some approach to equilibrium must have occurred somewhere in the Universe. Attempts have been made to account for the complete isotopic abundances of elements from chromium to nickel in terms of one equilibrium process. Such attempts have not been completely successful but one of the best fits is shown in fig. 37. As will be explained in Chapter 7, it is not now believed that the solar system iron peak elements have been produced in true equilibrium, but they certainly show some approximation to equilibrium.

The iron-helium phase change

To return to our discussion of the succession of nuclear reactions which occurs as the temperature of a gas is raised, by the time the temperature reaches 4×10^9 K the material is composed almost entirely of iron and its neighbours. As we have mentioned earlier, no further energy releasing nuclear fusion reactions can occur after

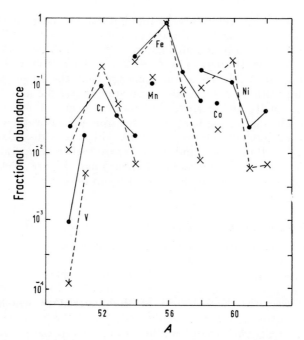

Fig. 37. The equilibrium process. The best fit of calculated results to the observed solar system iron peak abundances. Filled circles are observed values and crosses calculated values.

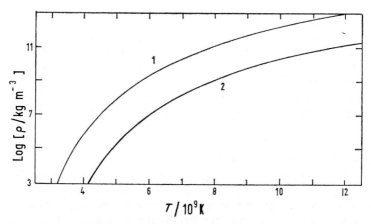

Fig. 38. The conversion of iron into helium and neutrons at high temperatures. In statistical equilibrium on curve 1 ^{56}Fe has been converted to 50 per cent iron and 50 per cent helium and neutrons. Curve 2 shows where the helium is 50 per cent converted into protons and neutrons.

this stage. It might, therefore, be thought that no change in chemical composition would occur if the temperature were raised still higher. This would not be true and an analogy will serve to show what would happen. At low temperatures a proton and electron will form a hydrogen atom which has a lower energy than the single particles. As the temperature rises, the hydrogen atom is still the most strongly bound state of the two particles but the hydrogen will eventually be ionized. The same thing happens to the iron nucleus as the temperature rises substantially above 4×10^9 K. First it is dissociated into α particles and neutrons and at an even higher temperature the α particles dissociate into protons and neutrons. Figure 38 shows the critical temperatures at which such break-up occurs.

Explosive thermonuclear reactions

In our discussion of thermonuclear reactions we have deliberately said very little about the astronomical objects in which the thermonuclear reactions occur as we wish to discuss those in Chapter 7. Our discussion has, however, been somewhat biased. We have supposed that the temperature of the material in which fusion reactions are occurring rises sufficiently slowly that the successive groups of reactions occur at the lowest temperature at which they are efficient. This discussion is appropriate to nuclear reactions in a slowly evolving star but there is also the possibility, which we have mentioned briefly at the end of the section on hydrogen burning, that material could be brought very rapidly to a temperature much higher than the minimum required for nuclear reactions to occur and that as a result the reactions occur explosively. Because all of the reaction cross-sections vary with temperature, the end product of such explosive burning of nuclei such as carbon, oxygen and silicon will be different from the end product of quasi-static burning. In Chapter 7 we shall see that both quasi-static and explosive thermonuclear reactions are believed to play a role in nucleosynthesis but the biggest change in the subject in recent years has been a shift of emphasis away from quasi-static reactions to explosive reactions.

The problem of lithium, beryllium and boron

As well as the proton-rich isotopes of the heavy elements, there is one further group of nuclei that is not produced by any of the main processes discussed in this chapter. These are the elements lithium, beryllium and boron and the heavy stable isotope of hydrogen, deuterium. The direct processes of hydrogen and helium burning produce only minimal amounts of these elements and they are very easily destroyed in stellar interiors, as they burn by nuclear reactions

90

at temperatures that are lower than those needed to burn hydrogen. Examples of these reactions are:

$$\left.\begin{array}{l} d(\text{p}, \gamma)^3\text{He}, \\ ^6\text{Li}(\text{p}, ^3\text{He})^4\text{He}, \\ ^7\text{Li}(\text{p}, \gamma)^8\text{Be} \rightarrow 2\,^4\text{He}, \\ ^9\text{Be}(\text{p}, ^4\text{He})^6\text{Li}(\text{p}, ^3\text{He})^4\text{He}, \\ ^{10}\text{B}(\text{p}, ^4\text{He})^7\text{Be}, \\ ^{11}\text{B}(\text{p}, \gamma)3\,^4\text{He}, \end{array}\right\} \qquad (5.40)$$

and the ^{7}Be produced in the fifth reaction is destroyed as in the PP chain.

These light elements are not very abundant but they must have been produced somehow. They are relatively more abundant in cosmic rays than in either the Earth or the Sun, although there are some stars which contain them in much higher amounts than either the Earth or the Sun. Their presence in cosmic rays has in Chapter 4 been attributed to break-up or *spallation* nuclear reactions caused by collisions of the heavier cosmic ray particles with interstellar protons. It seems quite possible that such spallation nuclear reactions have been the principal source of lithium, beryllium and boron throughout the Universe and we shall discuss this further in Chapter 7.

We have now described briefly the different processes which are believed to have contributed to the production of the chemical elements which we observe today. It will now be necessary in the next two chapters to discuss where and when these processes may have occurred in the Universe.

Summary of Chapter 5

Observations of the chemical composition of objects in the Universe strongly suggest the hypothesis that its initial composition contained only light elements and that the heavy elements have since been produced by nuclear reactions. Elements up to the neighbourhood of iron in the periodic table can be produced by nuclear fusion reactions which release energy, but an input of energy is required to initiate fusion reactions which produce heavier nuclei. Reactions between charged particles are very slow, when the nuclear charges are high, and it is believed that the heavy nuclei have been produced by neutron capture reactions.

Two processes of neutron capture have been proposed; the s-process in which the neutrons are captured so slowly that unstable nuclei decay before capturing another neutron and the r-process in which many neutrons can be captured before decay occurs. In an astrophysical context, it appears that the two processes should be essentially distinct. The r-process builds the most neutron rich isotopes, including all of the natural radioactive nuclei, whereas the s-process builds isotopes with relatively more protons and can build no isotope more massive than ^{209}Bi. Both processes produce abundance peaks close to neutron magic numbers. In the s-process, the product of neutron capture cross-section and isotopic abundance should be approximately constant and the solar system abundances of some isotopes agree with this prediction.

If material which is initially composed of light elements is gradually heated, a succession of nuclear fusion reactions occurs. At successive temperatures near to 10^7 K, 10^8 K, 5×10^8 K and 10^9 K, hydrogen burning, helium burning, carbon burning and oxygen burning take place. When the temperature exceeds 2×10^9 K a complicated chain of reactions, which has been given the name of silicon burning, leads to the production of the elements of the iron peak. This succession of fusion reactions may occur in a star, which is evolving slowly, but they can also occur explosively if material is suddenly raised to a temperature higher than the minimum required for the reaction to occur. It appears that a substantial fraction of nucleosynthesis may occur in explosive processes.

Two small groups of isotopes of low relative abundance are not produced by either neutron capture or thermonuclear fusion. These are the most proton-rich isotopes of the heavy elements, which may be produced by a process of fast proton capture and the elements lithium, beryllium and boron, which are probably produced by the breakdown of somewhat heavier nuclei by protons.

CHAPTER 6
cosmological element production

Introduction

IN previous chapters we have discussed observations of the chemical composition of the Universe and the processes by which that composition can have been produced. In this chapter and the next we are concerned with the evolution with time of the chemical composition of the Universe. In this chapter we are interested in what was the chemical composition before the first galaxies were formed; the study of the initial chemical composition of the Universe may be called *nucleogenesis* or *cosmological element production*. In Chapter 7 we discuss the manner in which the chemical composition of a galaxy changes during its life history. We shall study the chemical composition of our own Galaxy, because most observations refer to it: however it can be regarded as representative of other galaxies.

Cosmology is the study of the whole Universe rather than particular parts of it which is the domain of *astronomy*. The aim of a *cosmological theory* is to discuss the evolution with time of the large scale structure of the Universe and to calculate values of quantities such as the mean density and temperature of matter. A successful cosmological theory should also discuss the way in which smaller objects such as galaxies are formed. Cosmological element production implies the simultaneous synthesis of elements throughout the whole or a large part of the Universe. In a limiting case it may merely be the specification of the initial chemical composition of the Universe.

The simplest form of cosmological theory would say that the chemical elements were created as they are now. This would probably have been the answer before the discovery of natural radioactivity and the growth in knowledge about nuclear structure. The existence of radioactivity shows that some changes of chemical composition must have occurred since the radioactive elements were formed, and the study of stellar evolution has indicated that other nuclear transmutations have occurred in the lifetime of the Universe. We have seen that hydrogen and helium are overwhelmingly more abundant than other elements but that there are large variations in the proportion of heavier elements from star to star in the Galaxy. This suggests an investigation of the hypothesis that initially the Universe contained only light elements and that the heavier elements have all been produced during the lifetime of the Galaxy. In the present chapter we ask whether current observations of the large scale structure of the Universe can give any hints about its initial chemical composition. This sounds rather grandiose but we shall really be concerned with asking whether it is reasonable to suppose that initially the Universe was composed of hydrogen alone or whether it contained a mixture of both hydrogen *and* helium.

Observations of cosmological significance

We first list those observations which we believe to be of cosmological significance and which must therefore be taken into account in the construction of any cosmological theory. They include the following, which will be discussed in detail below:

- (i) The red-shifts of distant galaxies.
- (ii) The isotropic distribution of galaxies.
- (iii) The homogeneous distribution of *nearby* galaxies.
- (iv) The distribution of *distant* galaxies and radio sources.
- (v) The microwave background.
- (vi) The X-ray background.

Before explaining what is involved in each of these observations and trying to interpret them, one point must be stressed. The *laws of physics* as we understand them have been established by experiments in terrestrial laboratories and by observations in the neighbourhood of the Earth. In discussing observations of distant galaxies it is simplest to assume that the laws of physics established on Earth are true throughout the Universe, unless this leads to a contradiction. It is difficult to see what other assumption can be made. As light from distant objects has taken thousands of millions of years to reach us, we are assuming that the laws of physics are unchanging in time as well as in space.

Some theories have been proposed in which quantities such as Planck's constant, h, or the gravitational constant, G, vary with time. In discussing variations of these quantities, it is most sensible to discuss dimensionless numbers such as the ratio of the gravitational and electromagnetic forces between two particles. In the case of the proton and the electron, this is

$$R \equiv 4\pi\epsilon_0 G m_p m_e / e^2 \simeq 4 \times 10^{-40}. \qquad (6.1)$$

Some physicists believe that the gravitational force between two particles depends on the distribution of matter throughout the Universe and that, if the mean density of the Universe was higher in the past than it is now, the value of R would then have been higher.*

* In the Big-Bang theory to be described below, in which the density of the Universe does vary with time, $R^2N \approx 1$ at present, where N is the total number of particles in the *observable Universe*. If R does not change with time, it is a coincidence that we are alive when $R^2N \approx 1$.

This general belief that the gravitational constant is determined by the distribution of matter in the Universe has the name of *Mach's principle*, but there is at present no formulation of it which leads of a clear law of variation of R. For that reason we assume that the laws are unchanging. This assumption may be true but we should always remember that it need not be.

We can now discuss each of the observations listed above in turn:

(i) *The red-shifts of distant galaxies*

It is observed that spectral lines from distant galaxies are shifted towards the red end of the spectrum and that this red shift is approximately proportional to the distance of the galaxy from the Earth. The simplest interpretation of the red shift is that the distant galaxies are moving away from us* with the velocity of recession depending on distance and the red shift caused by the Doppler effect. We thus often speak of the *recession of the galaxies* or the *expansion of the Universe*. We have approximately

$$v = Hr, \tag{6.2}$$

where v is the velocity of recession, r is the distance and H is called Hubble's constant.† The definition of distance is itself not easy for distant galaxies; to a first approximation we will assume that it is what is known as *luminosity distance*, which is deduced by the assumption that galaxies which look structurally similar are the same size and that their apparent luminosity is thus a measure of distance. There is considerable uncertainty in the value of H but at present it seems likely that H^{-1} is between 4×10^{17} and 6×10^{17}s ($\sim 1 \cdot 5 \times 10^{10}$ yr).

(ii) *The isotropic distribution of galaxies*

It is possible to survey equal areas of the sky in different directions and to count how many galaxies are visible in each area. Provided the region being studied does not lie too close to the plane of the Galaxy (see fig. 39), it is found that these numbers are not very different; although differences exist they are not statistically significant and we say that the galaxies are distributed *isotropically*. There are smaller numbers visible in directions close to the plane of the Galaxy (this was named the *zone of avoidance*), but this is almost certainly

---

* The reason why this does not imply a geocentric Universe is discussed on page 100.

† This is named after Edwin Hubble, who first established the velocity-distance relation in the 1920's. His early work is described in his book *The Realm of the Nebulae* which is listed in the *Suggestions for further reading* on page 169. Because of difficulties in calibrating the distance scale, his original value of H^{-1} was only $1 \cdot 8 \times 10^9$ yr. At this time the *age of the Universe* appeared to be less than the age of the Earth and this caused difficulties for cosmological theories.

merely caused by absorption of their light by interstellar gas and dust in the Galaxy. As the distribution of galaxies looks isotropic *from the Earth* it appears that the Earth is in a special position in the Universe. A discussion of why this need not be so is given below on page 100.

Fig. 39. The zone of avoidance. Because of absorption by interstellar matter near the galactic plane, very few galaxies can be seen in directions between the two dashed lines (× is the position of the Sun).

(iii) *The homogeneous distribution of nearby galaxies*

If we now divide the galaxies in the various regions of the sky into groups at different distances, we can study the spatial distribution of the galaxies in each direction and can ask whether there is any uniformity in this distribution. Here we must be careful because in looking to large distances we are also looking into the remote past. It we assume that the results described in (i) above imply that the Universe is and has been expanding, galaxies should have been closer together in the past than they are now. When we observe very distant galaxies we are seeing them as they were a long time ago. If we could observe galaxies at sufficiently great distances, we would expect them to be nearer together than galaxies in our neighbourhood. In fact, we cannot observe ordinary galaxies at large enough distances for this effect to be significant. The characteristic time for the expansion of the Universe from (i) should be about 10^{10} yr and, if we are studying galaxies out to a distance of $1 - 2 \times 10^9$ light yr or less, which is about as far away as we can study galaxies in detail, the crowding together should not be significant.

Given that this effect is unimportant, we can now ask whether the galaxies we can observe are distributed approximately uniformly in space. Although there is a strong tendency for galaxies to be grouped in clusters, it appears on average that the relatively nearby galaxies are distributed uniformly in space and we describe this as a *homogeneous distribution.*

(iv) *The distribution of distant galaxies and radio sources*

Under this heading we are asking whether there *is* a bunching together of galaxies at large distances and therefore in the remote past. Unfortunately ordinary galaxies cannot be studied at great enough distances for the test to be made but observations of distant radio sources, including *radio galaxies*, are possible. For nearby radio sources, both radio and optical emission can be studied and different classes of radio sources can be identified. There are in addition many radio sources from which no optical emission can be observed and it is believed that they are more distant. If it is assumed that the distant radio sources fall into classes similar to the nearby radio sources, we can ask whether their space density is greater at large distances. This problem of the *radio source counts* has been a source of great controversy between M. Ryle and F. Hoyle amongst others in the past 15 years, but it is now generally (but not universally) agreed that the observations of Ryle and his colleagues have shown that either there is a larger density of radio sources at large distances or that the distant radio sources have distinctly different properties from the nearby sources. In either case we have evidence that the population of radio sources in a given region of space in the remote past is different from that in a similar region of space today. As a crowding together of sources is not proved, we do not at present have clear evidence for expansion from this observation.

The quasars

A central role in disputes about the radio source counts has been played by the objects known as *quasars*. These objects, of which the first were identified in 1963, are typically strong emitters of both optical and radio waves. Optically their apparent size is very small and they were first thought to be stars in the Galaxy. Then it was found that their spectra showed red shifts and, if these were interpreted according to the Hubble relation (6.2), they were the most distant objects known in the Universe. Unlike galaxies which show large red shifts, quasars cannot be identified with similar objects whose distance is known and there is no way of deducing their distances except through the Hubble relation. If they are at the distance predicted by (6.2), they are the brightest objects known in the Universe and their luminosity is produced in a very small volume. It has proved difficult to understand how the energy radiated by quasars is produced and this has led some workers to suggest that the red shifts are not cosmological in origin; one suggestion is that they are objects moving with a velocity close to that of light, which have been thrown out of a much nearer galaxy by an explosion. If they are nearer, their apparent brightness represents a much smaller total output of energy, but it is difficult to understand why no blue shifted objects which are moving towards us have been detected. The supporters of local quasar theories suggest that many of the radio sources which Ryle claims to have detected at large distances are quasars which are much nearer than he thinks, but his interpretation of the radio source counts still receives the greatest support.

(v) *The microwave background*

In 1965, A. E. Penzies and R. W. Wilson were working at Bell Telephone Laboratories on a microwave device and found that however hard they tried they could not remove some unwanted microwave noise from their circuits. Eventually they realised that the microwaves were coming from outer space. Subsequently the microwaves were detected by many other observers and it was shown that they were reaching the Earth *isotropically*. The isotropy of the microwave background radiation is much more definite than the isotropy of the distribution of galaxies and it appears to be accurate to a fraction of 1 per cent. The possible significance of the microwave background will be discussed in connection with the big bang cosmological theory below.

(vi) *The X-ray background*

More recently it has been shown that we are receiving a flux of X-rays from outer space. At present, the direction from which the X-rays are coming cannot be measured as accurately as in the case of microwaves but the X-ray flux is certainly relatively isotropic and it may be really isotropic.

Cosmological theories

We can now discuss cosmological theories and ask how the above observations fit into them. In attempting to produce a theoretical description of the entire Universe, it is natural to try to obtain the simplest description which accommodates all of the significant observations. Thus the cosmological theories which we will describe are very simple but this does not imply a conviction that the Universe really is simple; it will be assumed to be simple until observations demonstrate that this assumption is invalid.

Oscillating and big-bang cosmologies

If we accept the interpretation of observation (i) given above, the Universe is at present expanding. Other interpretations of the red shift have been proposed but none of them has received many supporters. The red shift of galaxies certainly increase with their distance from us so that any satisfactory explanation of the red shift must give this property. It *has* been proposed that photons are automatically shifted to the red by travelling large distances; this would give the observations without any expansion but it is regarded as an arbitrary assumption as we have no independent experimental evidence, whereas the Doppler explanation is well tested. If the Universe is expanding, we can assume either that it has *always* been expanding or that at certain times in the past it has been contracting.

This leads to two simple types of cosmological theory; the *big-bang cosmologies* and the *oscillating cosmologies*. These are illustrated in fig. 40. In the big-bang theories the Universe had an origin at some time t_0 in the past whilst the oscillating cosmology has an infinite past life history.

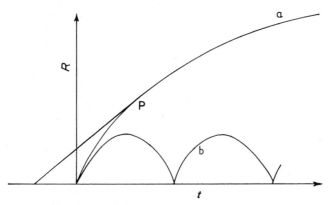

Fig. 40. The radius, R, of the volume occupied by a given mass of matter is shown as a function of time, t, for (a) a big-bang cosmology and (b) an oscillating cosmology. If P is the present epoch and if the big-bang theory is valid, the tangent to the (R,t) curve at P intersects the t axis at a negative value of t, showing that the age of the Universe deduced from the present rate of expansion is greater than the true age.

The steady state cosmology

In the big-bang theory galaxies would have been closer together in the past and this tends to be supported by observation (iv) above, on page 97. The reason why Hoyle has been concerned with this observation is that in 1948 he and H. Bondi and T. Gold suggested an alternative theory of the expanding Universe to the big-bang theory. They suggested that the *average* properties of the Universe were not changing with time. Clearly the properties in any small locality are changing as stars are born, evolve and die but their suggestion was that the total content of a volume containing many galaxies, but very small compared to the entire Universe, would be unchanging in time. As they also agreed that the Universe was expanding, this implied that there must be *continuous creation of matter* to make good the gaps left by the expansion and from this matter new galaxies and stars could be formed. This theory is known as the *steady state* theory*. According to it the Universe has an infinite past as in the oscillating theory; if we imagine it run backwards in time, matter disappears so that the average density of matter in any volume of space does not

* One initial attraction of the steady state theory was that in an unchanging Universe it is necessary to have unchanging laws of physics.

increase. In such a steady state Universe the average density of galaxies should be the same at all points at all times and the simple steady state theory cannot be true if the observation described in (iv) above is correct.

Interpretation of observations of isotropy

All of these three simple theories, (i) big-bang, (ii) oscillating, (iii) steady state, have in common that the Universe is *homogeneous and isotropic* at any time. We have, of course, only observed that the Universe appears to be isotropic as observed from the Earth and it might be thought that this implies that the Earth is in a privileged position and that the Universe could not appear isotropic from another position. This is not true, but to explain it in detail we should have to discuss the *general theory of relativity* which is outside the scope of this book*. General relativity attributes properties to space and time which are outside our ordinary experience. In Euclidean three dimensional space, which we normally study, a finite distribution of galaxies cannot appear isotropic from more than one point inside the distribution but in the *curved space-time* of general relativity it is, for example, possible to have a distribution which is simultaneously *finite and unbounded* and which looks isotropic from every point. This possibility for a two dimensional curved space can be illustrated; a mathematical description can be given of curved spaces of a higher number of dimensions, even if we cannot visualize them in our three dimensional framework.

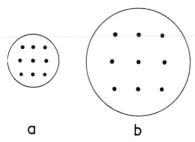

a   b

Fig. 41. A two-dimensional model of the recession of the galaxies. If dots are equally spaced on the surface of a sphere as in a, they continue to be equally spaced when the sphere is expanded in b. In addition each dot sees the others receding from it with a velocity proportional to distance.

An example of a two dimensional curved space is the surface of a sphere and this is both finite and unbounded. In order to have these properties it has to be embedded in a space of three dimensions. If the surface of the sphere is covered uniformly with spots, from any one of them the others will appear to be uniformly distributed in the

---

* An introductory book on General Relativity is listed in the suggestions for further reading on page 169.

two dimensional curved space.   Moreover, if the sphere as a whole is expanded, the spots will move apart with a relative velocity which is proportional to distance (fig. 41), so that we obtain a two dimensional analogue of the recession of the galaxies and Hubble's law.

If we now suppose that the Earth is not in a privileged position and that the Universe appears to be to a high degree isotropic from all points within it, we can see the full significance of the observations of the microwave and X-ray backgrounds.   We expect these to exist at all points so that the Universe is filled with microwaves and X-rays.

Cosmological element production

Steady state theory.   In the steady state theory we require continuous creation of matter but there is no clear indication of what form that matter should take.   In fact, we could adopt the view that the initially created matter has an unknown composition which is also the initial composition of galaxies.   We could then attempt to calculate how much the composition has been changed by nuclear reactions in stars and other objects in the lifetime of the Galaxy.   Assuming that the present composition is known, we could then in principle deduce the initial composition and this could be assumed to be the composition of the matter which is continuously created.   In the following chapter, we *shall* try to relate the present composition of the Galaxy to its initial composition and we shall see that there are many uncertainties in the procedure.

In fact, this attitude has not been taken.   The steady state theory was proposed as a simple theory and its supporters have always felt that it is a necessary feature of the theory that the composition of the created matter is simple.   The only two serious proposals that have been made are the creation of neutrons and the creation of hydrogen. The first of these is untenable.   The neutrons would soon decay by the reaction

$$n \rightarrow p + e^- + \bar{\nu} \qquad (6.3)$$

and the energy released in this reaction would heat the intergalactic gas to a temperature of about 10^9 K.   Such a hot intergalactic gas would in turn radiate gamma rays with an intensity which should already have been detected.   Protagonists of the theory have thus supposed that the initially created matter is hydrogen and that all other elements must have been produced by nuclear reactions in stars or other objects.   In Chapter 7, we shall consider whether the initial composition of our Galaxy could have been pure hydrogen.

The big-bang theory.   Although we refer to the big-bang theory, there is not a uniquely defined theory.   The Universe is supposed to have come into being at some definite time in the past, or at least to have received its present form then.   Immediately after the moment of creation the Universe was very dense.   In the most popular version of the theory the Universe is also supposed to have been very hot. In the early stages of such a hot big bang, the matter in the Universe is composed of protons, neutrons, electrons, positrons, neutrinos, antineutrinos and other elementary particles. As the Universe expands it cools and, when the temperature drops below about 10^{10} K, neutrons and protons start to combine to form deuterons through the reaction

$$n + p \rightarrow d + \gamma. \tag{6.4}$$

The reaction would also have occurred at higher temperatures but the reverse reaction would have occurred equally rapidly, so that there would be no net build-up of deuterium. The formation of deuterium is quickly followed by the production of helium and detailed calculations show that, by the time the Universe has cooled to the extent that no further nuclear reactions are possible, the chemical composition would be essentially a mixture of hydrogen and helium. To make the theory well-defined, it is necessary to specify the matter density at some prescribed temperature (10^{10} K, say). If this is done, the actual predictions of the theory are shown in fig. 42. In this theory, the early nuclear reactions are over in about 1 hr so that it is reasonable to say that the initially created matter is a mixture of hydrogen and helium.

Oscillating theory. If we suppose that there have been an infinite number of oscillations in the past, it seems that, as far as chemical

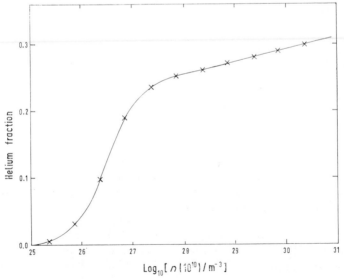

Fig. 42. Helium production in the hot big bang. The fraction by mass in the form of helium after the initial stages of the expansion is shown as a function of the nucleon density [$n(10^{10})$] when the temperature is 10^{10}K.

composition is concerned, the theory must have much in common with the hot big-bang theory. In any one oscillation of the Universe, nuclear reactions occur in stars leading to the production of heavy elements including small amounts of iron. Once iron is produced, it is difficult to destroy it and it would appear that the amount of iron would be increased in each oscillation. After an infinite number of oscillations, it might be expected that the material would be almost pure iron. This is certainly not in agreement with observation. The theory can be reconciled with observation only if at the minimum radius of each oscillation the temperature and density of the Universe are so high that the heavy elements formed in the earlier stage of the oscillation are dissociated. If the Universe does become that hot in each oscillation, the early stages of the next oscillation closely resemble a hot big bang.

The microwave background

When Gamow first propounded the hot big-bang theory and the $\alpha\beta\gamma$ theory of element formation which was mentioned in Chapter 5, he noted that the Universe would have been filled with radiation in its early stages. Unless that radiation has been absorbed since then, it should be around today but because of the expansion of the Universe it should appear at very much longer wavelengths because of the red shift. At very high temperatures the radiation would have a Planck distribution

$$B_\nu(T) = \frac{2h\nu^3/c^2}{\exp(h\nu/kT) - 1}. \tag{6.5}$$

It can be shown that, as the Universe expands, the radiation retains a Planck spectrum but the temperature becomes very much lower. Gamow estimated that the Universe should today be filled with radiation at a temperature of about 20 K, which has its maximum intensity in the microwave region of the spectrum, but at that time it did not seem possible that the radiation could be observed because microwave techniques were not sufficiently developed.

This prediction was essentially forgotten until 1965, when Penzies and Wilson discovered the isotropic microwave background radiation. When this had been observed at several wavelengths, it appeared possible that the radiation could have the Planck spectrum at a temperature of about 3 K*. If the background radiation were indeed a relict of the big bang, it should be possible to see whether the helium content of the Universe is also consistent with the theory. Figure 42 gives the initial helium content in terms of the density

---

* This differs from Gamow's value mainly because Gamow assumed that the Universe could contain neutrons and photons alone initially and he neglected all of the other elementary particles.

when the temperature of both matter and radiation was 10^{10} K. The big-bang theory moreover relates the density at 10^{10} K to the density when the radiation temperature is 3 K. The present mean density of the Universe is somewhat uncertain but it is believed to lie in the range 10^{-1} to 10 particles per cubic metre. If we use these values, we find that the predicted initial helium content is between 23 per cent and 28 per cent by mass.

This observation immediately gave great popularity to the big-bang theory because most objects in the Universe, for which a helium content is known, contain 25 per cent helium or more by mass. In addition, as we shall see in the next chapter, it is rather difficult to believe that all of this helium can have been produced in stars. Although the theory became attractive, it must survive two crucial tests as observations accumulate. In the first place it is necessary that all objects, except possibly a few which have had a particularly peculiar evolution, should have a higher helium content than the minimum amount produced in the big bang. In the second place the black body nature of the microwave radiation must be confirmed.

At present there is uncertainty with regard to each of these observations. Several times during the past five years it has been suggested that there exist some old stars which have evolved normally and which

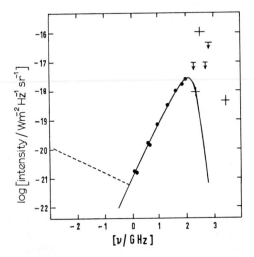

Fig. 43. Observations of the microwave background radiation. The filled circles are ground based observations which all fit the low frequency end of the solid curve which is the black body curve for 2·7 K. The crosses are some of the observations made from above the atmosphere; these results are very irregular and different groups of observers do not always get results which agree. The arrows denote observations which are only upper limits. The dashed line denotes radiation from known radio sources at very low frequencies.

have an extremely low helium content. If these observations were confirmed, it would be difficult to support the hot big-bang theory, but at present none of them has been universally accepted. In Chapter 3 we have explained that it is sometimes very difficult to deduce reliable abundances from stellar spectra and this is true for these stars. There is also an observation that helium absorption lines are weak in the light from some quasars. Again, this could imply that the Universe had a low initial helium content, but the interpretation of these observations is also disputed.

The present state of the observations of the microwave radiation is shown in fig. 43. The observations from ground-based telescopes, which are at the longer wavelengths, are in agreement with the black body spectrum but there have in addition been some observations by rocket-borne instruments, which indicate that the spectrum is not black body at shorter wavelengths. It has not yet been possible to study the entire wavelength range and further observations are awaited. What may be significant is that no observations give an intensity that lies below the black body curve and it is possible that we are observing black body radiation from the early Universe on which is superimposed radiation from sources which have come into being since then; certainly the black body curve is swamped at very long wavelengths by the emission from known radio galaxies.

The formation of galaxies

In the simple theories which we have been describing the Universe is supposed to be homogeneous; in particular in the steady state theory a uniform creation rate of matter is assumed whilst in the big-bang theory it is assumed that the Universe genuinely is homogeneous in the earliest stages, so that the chemical composition at the end of these initial stages is also homogeneous. It is, of course, a clear fact of observation that, whilst the Universe is homogeneous on the average, it is not in detail homogeneous. In particular galaxies exist. We do not at present know whether there exists a substantial amount of intergalactic matter but, whether or not there does, the galaxies introduce a distinct irregularity into the mass distribution. In the simplest cosmological theories, it is assumed that the galaxies are formed as a result of the contraction of self-gravitating condensations which arise in intergalactic space; the galaxies are supposed later to form into stars through a similar process.

A simplified description of how self-gravitating condensations arise in a uniform medium which is not expanding is as follows. In such a uniform medium small density irregularities will arise spontaneously. The irregularities will cause sound waves to propagate through the medium and these will smooth out the irregularities unless the self-gravitation of the region causes it to

105

contract more rapidly than the sound waves can dissipate it. Self-gravitation makes the region contract significantly in a time $(G\rho)^{-\frac{1}{2}}$ where ρ is the density; this is like the dynamical timescale for the collapse of a star and it can be obtained from dimensional arguments. If the density irregularity has length scale ℓ and the sound speed is c_S, it will take sound a time ℓ/c_S to cross the irregularity. If this time exceeds $(G\rho)^{-\frac{1}{2}}$ the irregularity can become more pronounced and collapse can occur. Fluctuations of all length scales will arise spontaneously but only those of large dimension can lead to collapse. This process is known as *Jeans's instability* after the astronomer who first studied it.

The above description does not quite apply to the formation of galaxies, because in that case collapse must occur against the background expansion of the Universe as a whole and the calculation must be performed in the curved space-time of relativity. The discussion can be generalized. If the Universe were initially homogeneous, the galactic masses would have to arise through accidental irregularities in the initial smooth Universe. This means that we must ask whether such small irregularities could grow to form galaxies in a time of order H^{-1}. There have in the past been great difficulties in demonstrating that galaxies could form in such a time from initial infinitesimal irregularities, although there would be no difficulty if the initial irregularities were larger. This means that there has been speculation that the Universe has never been strictly homogeneous but that it has possessed structure throughout its life history. There is some reluctance to adopt this view because it really amounts to saying that galaxies exist now because they have always been in existence. If the Universe has always been inhomogeneous, the chemical composition need not be regular; this could be particularly true if some modified version of the big-bang theory is correct.

Intergalactic matter

One very important observation would be the demonstration of the existence of an *intergalactic medium* with a significant density. It seems just possible that, if an intergalactic medium exists, something could be learnt about its chemical composition. A discovery that it was deficient in helium would be most important. It would then suggest that the Universe has always been significantly inhomogeneous and that helium production has occurred in the protogalaxies but not in the low density regions between the galaxies. The amount of helium produced might then be expected to depend on the mass of the galaxy and this suggests that it is worthwhile to try to discover whether there is any observational evidence for a variation of helium content from galaxy to galaxy.

Summary of Chapter 6

The observations of the distribution of galaxies and radio sources and the X-ray and microwave backgrounds suggest that there is a considerable amount of regularity in the Universe. The observations of the red shifts of distant galaxies suggest that the Universe is expanding. Several simple models of the Universe have been suggested and studied. If three crucial observations are confirmed by future work, the hot big-bang theory looks very attractive as a *first approximation* to the description of the Universe. These observations are:

(a) Radio sources were either much closer together or much stronger emitters of radio waves in the past.

(b) There is an isotropic background microwave radiation which is black body at a temperature of about 3 K.

(c) All objects in the Universe have a helium content of about 25 per cent or more by mass.

No one of these observations is clearly established at the moment. The discovery of any significant group of objects with a helium content lower than 25 per cent or the demonstration that the microwave background is not black body would invalidate the theory. If the value of 20 per cent for the solar helium content, which is mentioned in Chapter 3, is confirmed, it is already difficult to accept the theory in its simplest form.

If the observation (a) above is clearly established, the steady state theory is invalid. Even if (a) is discounted, the steady state theory has some difficulty as it must explain the microwave background, which can no longer be relict radiation, and it may have to explain a high helium content, which can no longer be primaeval, unless we assume continuous creation of a *mixture* of hydrogen and helium. There are now few, if any, supporters of the simplest version of the steady state theory.

Simple cosmological theories have been studied first because that is the easiest thing to do but that need not imply a conviction that the Universe really is simple. Various observations and theoretical problems, including that of galaxy formation, suggest that the Universe may always have possessed serious inhomogeneities. This leads to the introduction of inhomogeneous big-bang cosmological theories. As supporters of continuous creation are now being forced to replace the uniform steady state theory by irregular creation, the distinction between the two types of theory is perhaps not as great as it was.

CHAPTER 7

element production in the galaxy

Introduction

WE now turn to what is perhaps the most difficult and uncertain part of our subject. This is concerned with the extent to which the chemical composition of the Galaxy has changed since it was formed and what objects are the principal centres of nucleosynthesis. Although the title of this chapter refers to element production in the Galaxy, we shall regard the Galaxy as a representative of all galaxies; we expect similar processes of element formation to have occurred in most, if not all, of them, but the detailed observations, which can be discussed, are mainly confined to our Galaxy.

We have discussed already in Chapter 3 what evidence there is for variations in chemical composition between individual stars in the Galaxy. We have seen that there is evidence that some very old stars have only a very small amount of elements heavier than helium and that it is also possible that stellar chemical composition depends on the position of the star in the Galaxy. It has been stated in Chapter 3 that very low abundances of heavy elements *only* occur in the oldest stars. It thus appears that, if the initial heavy element abundance in the Galaxy were uniform in space, it must have been extremely small, but that a large fraction of the presently observed heavy elements must then have been produced very rapidly compared to the total age of the Galaxy. There is also the tendency, which we mentioned in Chapter 3, for the heavy element content of young stars to be higher near the centre of the Galaxy than further out and certainly there is lack of uniformity in the chemical composition of stars of a given age. This indicates that the rate of production of heavy elements has not been uniform in space and that there may have been preferential production of heavy elements in the central regions of the Galaxy.

As we have already mentioned in the preceding chapter, the evidence about the abundance of helium is unclear. It is possible that all objects contain a large amount of helium, which was present in the Galaxy when it was formed, but it is certainly necessary to discuss the processes by which helium is produced and to try to decide whether all of the helium which is observed today could have been produced within the Galaxy.

One feature of the observations of the abundances of heavy elements is that, although there are some stars which have what are

108

described as peculiar chemical compositions, the *mix* of heavy elements is rather similar in most stars although the *total amount* is very variable. This has led to suggestions that one type of object might have been responsible for the production of most of the heavy elements. The argument for this point of view is that, if significant amounts of heavy elements are produced in a large number of different types of source, the mix of heavy elements might be expected to be much more variable. Although this argument is attractive, it is not completely convincing; if a very large number of sources of heavy elements contribute to the chemical composition of an individual star, the mix of heavy elements in different stars may be quite similar, even if individual sources of heavy elements vary greatly.

Two possible classes of object may have been responsible for most of the galactic nucleosynthesis. These are:

(i) stars:

(ii) some objects more massive than stars.

Because stars have been studied most, we first discuss nucleosynthesis in stars.

Nucleosynthesis in stars

In a discussion of nucleosynthesis in stars, there are at least four aspects to the problem.

First, we must study the evolution of stars of a wide range of masses and discover what nucleosynthesis will occur in them during their life history. We believe that a succession of nuclear fusion reactions which we have described in Chapter 5 occurs in stars above a minimum critical mass. We also believe that, at some stages of stellar evolution, neutron releasing reactions may occur, so that some of the neutron capture reactions which have also been described in Chapter 5, can then produce nuclei heavier than iron.

Second, we must study the way in which stars lose mass and expel synthesised matter into the interstellar medium.

It might be thought that the variation in the amount of heavy elements from star to star reflects the degree of nucleosynthesis that has occurred in the stars themselves. This is certainly not correct in general. We have good reasons for believing that main sequence stars are stars which are at an early stage of their evolution in which no significant nuclear reactions have occurred. In that case, the observed chemical composition of a main sequence star is essentially its original chemical composition and it is equivalently the chemical composition of the interstellar gas out of which the star was formed. In support of this view, there is good agreement between the chemical composition of the youngest main sequence stars and the present composition of the interstellar gas. We believe that a study of the variation of chemical composition with *age* and *place of origin* of main sequence stars can give information about the gradual change of the chemical composition of the interstellar medium in different parts of the Galaxy.

We believe that, if heavy elements are produced in stars, it is essential for them to be expelled from the star in which they have been produced, so that they can be used in a new generation of stars. That means we need to know when in its life history a star is likely to become unstable and to lose mass to interstellar space. We need to know when this instability will occur, how much mass will be lost and whether nuclear reactions causing important modifications in the chemical composition of the ejected mass occur during the process of ejection. We shall see that this leads to a discussion of the relative roles of *thermal processes* and *explosive processes* in nucleosynthesis; If, for example, the r-process of neutron capture occurs in a star, it must be an explosive process associated with the ejection of matter, because only in an explosion can neutrons be made available in high enough intensity for an r-process. In discussing mass loss from stars we shall be interested in what fraction of the mass of a star can be removed by explosion. This tells us how much of the total element production can be regarded as *useful production* and how much is *locked up* in the remnants of exploded stars.

Third, what happens to the material after it has been ejected from stars? For example, can we expect the material ejected from many stars distributed through the Galaxy to lead to an interstellar medium of approximately uniform chemical composition or should we expect the chemical composition of the interstellar medium to be seriously non-uniform? We are also interested in the efficiency with which interstellar matter is reconverted into stars.

Fourth, we are concerned with the number of stars which have played an important role in nucleosynthesis and how that number has varied as a function of both position and time. We have some idea of the *total* amount of heavy elements in the Galaxy at the present time. The total mass of the Galaxy is probably in the range $10^{11} M_\odot$ to $2 \times 10^{11} M_\odot$ and the mass of heavy elements in visible stars and the interstellar medium is probably of order $2 \times 10^9 M_\odot$ or greater. This is an estimate of what we have above called the useful production of heavy elements; there will be an additional quantity of heavy elements in the invisible remnants of exploded stars. If we think that we have identified the stars which play a major role in nucleosynthesis, have there been enough of them in the past life history of the Galaxy to give a useful production of heavy elements of order $2 \times 10^9 M_\odot$?

We discuss all of these aspects of stellar nucleosynthesis to the extent to which progress has at present been made. It must be stressed again that this work is still very incomplete and that no very definite conclusions can be drawn at present, but it *is* possible to show how work is proceeding. Subsequently, we discuss briefly the possibility that objects more massive than ordinary stars may have played a significant role in nucleosynthesis.

7.1 *Fusion reactions and stellar evolution*

We consider first the way in which the chemical compositions of stars change due to nuclear fusion reactions as they evolve. This has been described in some detail in the companion book, *The Stars: their structure and evolution*, and we summarize the conclusions without giving the detailed arguments.

Most of the energy which stars radiate is released by the thermo-nuclear reactions, which have been described in Chapter 5. As a star evolves, its central temperature increases for a time and a succession of nuclear fusion reactions ensues. How much element production occurs in a particular star depends on how long its internal temperature continues to rise. The knowledge that we have obtained about this from calculations of stellar evolution is summarized in Table 13. What we can say briefly is that the maximum internal temperature reached during stellar evolution is highest for the most massive stars. Thus the massive stars are the objects which possess the greatest *potential* for nucleosynthesis. This is true for a second reason. Massive stars are observed to radiate energy at a much higher rate than low mass stars and, whereas the stellar luminosity increases as a high power of the mass (except for the most massive stars), the maximum energy which can be released by nuclear fusion reactions in a star of given initial chemical composition only scales linearly with the mass. This means that massive stars complete their life history much more rapidly than low mass stars.

Mass/$M_\odot$	Nuclear reactions
0·08	None
0·3	H burning
0·7	H, He burning
5·0	H, He, C burning
30·0	All fusion reactions releasing energy

Table 13. The nuclear reactions which are believed to occur during normal evolution of stars of different masses are shown; the masses are only approximate and further reactions may occur if a star explodes.

The mass-luminosity relation for main sequence stars is shown in fig. 44, and the main sequence lifetime of stars of different masses is shown in Table 14. The main sequence lifetime of a star is comparable with the total life of the star, because most of the energy from nuclear fusion reactions is released in hydrogen burning. Stars less massive than the Sun have such a long main sequence lifetime that, even if they are almost as old as the Galaxy, they can have played no role in useful nucleosynthesis. Some nuclear reactions will have occurred within them but the processed material will still be deep in their interiors. In contrast, the whole sequence of nuclear fusion reactions can have occurred in massive stars which produce heavy elements very rapidly and there can have been several or even many generations of massive stars, which have contributed to the present chemical composition of the interstellar medium.

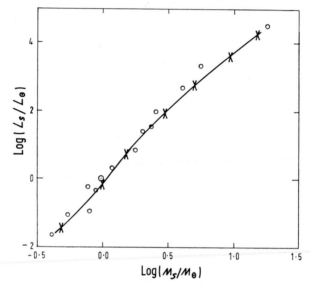

Fig. 44. The mass-luminosity relation for main sequence stars. The crosses and the solid line are theoretical results. The open circles and ⊙ (the Sun) are observational points.

$M/M_\odot$	Lifetime	$M/M_\odot$	Lifetime
15·0	$1·0 \times 10^7$	2·25	$5·0 \times 10^8$
9·0	$2·2 \times 10^7$	1·5	$1·7 \times 10^9$
5·0	$6·8 \times 10^7$	1·25	$3·0 \times 10^9$
3·0	$2·3 \times 10^8$	1·0	$8·2 \times 10^9$

Table 14. Main sequence lifetimes (in years) for stars of different masses.

Chemical composition of evolved stars

Detailed calculations of stellar evolution do not yet cover the complete life history of any star but for relatively massive stars a large fraction of their life history has been followed and we can hope to generalize the results a little. Consider the results of the calculations of the evolution of stars of 5 $M_\odot$ through hydrogen and helium burning to the onset of carbon burning. Figure 45 shows the chemical composition of such a star at this stage in its evolution according to the calculations of one group of investigators. Other calculations give results which differ in detail but agree on one feature of this figure, which will be very important later and which should be noted. There is a relatively thin belt of unburnt helium. Although a large amount of hydrogen has been converted into helium, much of this helium has been burnt into carbon. This seems to be a rather general result; however much helium has been produced by hydrogen burning, the total mass remaining as helium at any stage in the star's evolution seems unlikely to be much greater than 10 per cent of the mass of the star. This is very important, if all of the helium we observe today is not primaeval but has been produced in stars. If most stars today contain at least 25 per cent helium, but the helium has been produced in stars which can convert no more than 10 per cent of their mass into useful helium, *there must have been several generations of stars producing helium*.

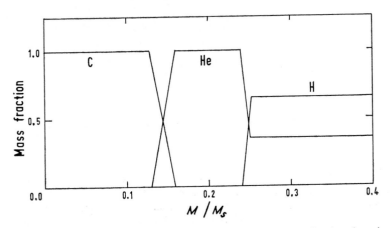

Fig. 45. The chemical composition of 5 $M_\odot$ star at the onset of carbon burning. The fraction by mass in the forms of C, He and H is shown for the inner region of the star.

Although no detailed calculations have yet been made of the complete structure of highly evolved stars including nuclear reactions such as oxygen burning and silicon burning, the chemical composition of a highly evolved star might appear schematically as shown in fig. 46.

Most of the calculations of stellar evolution, including those described above, have assumed that the stars evolve at *constant mass*. We shall shortly discuss processes of *catastrophic mass loss* which end the normal evolution of a star but there is also the possibility of

113

non-catastrophic mass loss at an early stage in a star's evolution. Such mass loss might not significantly alter the star's properties while it is occurring and it might be difficult to detect it observationally or to predict it theoretically. Nevertheless, it might have an important effect on the subsequent evolution of the star, because it alters the total mass of the star and leads to a different distribution of chemical composition through the mass of the star.

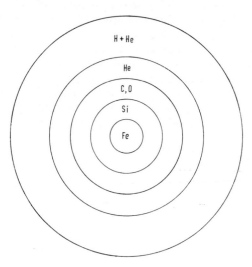

Fig. 46. Chemical composition of a highly evolved massive star.

7.2 *Mass loss from stars*

Before discussing the general problem of mass loss from stars, it is important to ask what is the final stage in the evolution of a star. It might be thought that, after a star has passed through the succession of nuclear fusion reactions listed in Chapter 5, it would gradually cool down and die as it continued to radiate energy into space. However, as has been explained in *The Stars*, only stars less massive than about $2 M_\odot$ can pass through a phase of nuclear evolution, reach a maximum temperature and then *cool down and die quietly*. Such stars can finish their lives either as *white dwarfs* (mass less than about $1 \cdot 2 M_\odot$) or as *neutron stars*. In the case of more massive stars, they can only die quietly if some instability or succession of instabilities causes their mass to fall below $2 M_\odot$.

In fact, it used to be asserted that such massive stars must lose all of their mass in excess of the maximum mass of white dwarfs* and

* At that time the possible existence of neutron stars was not fully accepted but it is now believed that neutron stars do exist and that *pulsars* are rotating neutron stars (see page 122).

this led to very favourable estimates of the amount of useful nucleo-synthesis which could occur in massive stars. More recently it has been realised that there is nothing in the presently accepted laws of physics which requires these stars to lose mass. If it is assumed that special relativity is valid in the sense that no signal speed can exceed the speed of light, that the force of gravitation is attractive for arbitrarily small separations between particles and that no new repulsive force, of which we have no present knowledge, dominates gravity at very short ranges, such objects collapse to a state of infinite density. If this happens, they become objects whose existence has not been conclusively demonstrated, which are variously known as *black holes* and *collapsars*. According to current theory, such objects would only interact with the remainder of the Universe through their gravitational force and attempts are being made to detect their presence as, for example, in the invisible component of a binary star system.

Whilst it was firmly believed that all massive stars must lose mass until their mass fell below $1 \cdot 2 \ M_\odot$, the study of nucleosynthesis was not too dependent on a detailed and accurate theory of mass loss from stars. Now that we can no longer assume that mass loss *must* occur, we must have a theoretical description of mass loss or observational evidence for it. We now consider this evidence.

Supernovae

In discussing mass loss we turn first to the type of object which is a popular favourite both as the main source of the heavy elements and as the origin of the cosmic rays; that is the supernovae. These are stars which have a very sudden increase in brightness followed by a less rapid decline; light curves of two supernovae are shown in figs. 47 and 48. At maximum brightness a supernova may give out as much visible radiation as an entire galaxy of ordinary stars. From the Doppler shifts of some lines in the supernova spectrum, matter is observed to be flowing outwards from supernovae at speeds of order $10^7 \ m \ s^{-1}$ and quite large amounts of mass are ejected in the explosions. If supernovae are highly evolved stars, it is possible that a considerable amount of processed material is ejected into space. We shall have more to say about this shortly.

Perhaps the best known supernova is the one observed by the Chinese astronomers in 1054 AD, which produced the Crab nebula (see fig. 25 of page 58). The Crab nebula contains filaments of gas which are observed to be expanding away from a central point and only a small change in expansion speed in the last 900 yr would be necessary to trace their origin back to the explosion of 1054. Radio astronomers have discovered that there is considerable radio emission from the Crab nebula, and from the expanding filaments of gas which form the remnants of other supernova explosions. Radio astronomers are able to identify supernova remnants from their characteristic radio properties, even when the explosion occurred too long ago for us to have any direct knowledge of it.

As has already been mentioned in Chapter 4, this radio emission is most readily understood if it is produced by electrons moving in spiral paths in a magnetic field. Accelerated charged particles radiate

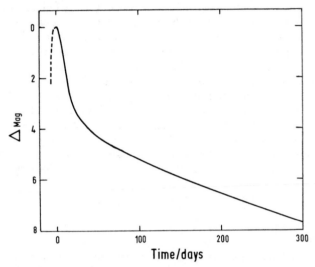

Fig. 47. The light curve of a supernova of Type I. The rise to maximum light shown by the dashed section is not usually observed.

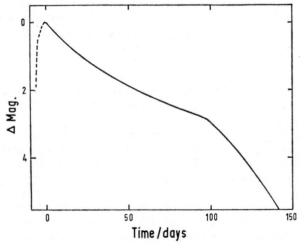

Fig. 48. The light curve of a supernova of Type II.

116

electromagnetic waves and, if the acceleration is caused by a magnetic field, the radiation is known as synchrotron radiation because it is similar to that produced in the particle accelerator known as a synchrotron. To produce the synchrotron radiation from supernova remnants, the electrons must be moving with speeds very close to that of light. The existence of these relativistic electrons is one reason why it is thought that supernovae may be the site of the origin of cosmic rays; if electrons can be accelerated to relativistic speeds in a supernova explosion, it is *possible* that the same is true of protons and other atomic nuclei. Another reason why supernovae are associated with cosmic rays is that, apart from explosions which are observed to occur in the central regions of some galaxies, the supernova is the most violent event which we observe.

Mass loss from supernovae

Given that there is a significant loss of mass in a supernova explosion, we are interested in whether the mass lost represents most of the mass of the supernova or whether, although the lost mass may be substantial in terms of the solar mass, it may be a relatively small fraction of the mass of the star. There are two reasons for being interested in this. The heavy elements ejected from a star form what has previously been called the useful production of heavy elements. If most of the mass of the star survives the explosion, it will preferentially contain material which has been processed by a succession of nuclear fusion reactions and this material will not be available to be incorporated into future generations of stars. The remnant will become a white dwarf, neutron star or a gravitationally collapsed object. The second point is that, if only the *outermost* layers of a supernova are ejected, they are likely to be composed of light elements such as hydrogen and helium at the time of the explosion. If supernovae lose only a relatively small fraction of their mass, the origin of the heavy elements can only be placed in the supernovae, if the heavy elements are produced explosively by nuclear reactions which occur *at the time of the outburst* rather than by thermonuclear reactions in the earlier stages of stellar evolution.

It is natural at this stage to ask about the observational evidence for the amount of mass lost. Unfortunately, this is not very clear. Only three or four supernovae have been observed in our Galaxy in the past thousand years and the last was that observed by Kepler in 1604. Many more supernovae have been observed in nearby galaxies in recent years. It is possible to observe the light curves and spectra of these supernovae and to show that mass is leaving them at high speed, but it is much more difficult to estimate how much mass is being lost and determine the chemical composition of the ejected mass. Although supernovae are intrinsically very bright, most of

117

those that have been studied have been so distant that they have appeared very faint. A considerably greater amount of information could be obtained if a supernova were to occur within our Galaxy and in a direction in which its light was not dimmed by interstellar absorption.

In addition we have no direct knowledge of the initial masses of the stars which become supernovae. Supernovae are usually divided into the two groups, Type I and Type II, whose light curves have been illustrated in figs. 47 and 48, although there are some that cannot be fitted into either of these groups. Type I supernovae are believed to be of *low mass* because they occur in stellar systems which mainly contain low mass stars and Type II supernovae are believed to be *massive* stars, because they occur in regions of galaxies which also contain massive main sequence stars, but these general ideas are no real substitute for a reliable mass determination.

A fair statement about the present observations of supernovae is perhaps that there is no *clear* evidence that *most* of the mass of a star is ejected in an explosion or that the ejected material is *mainly* in the form of heavy elements. Very few supernovae have been able to be studied in detail, so that it is not certain that few that have been studied are *typical* of the whole group. Astronomers would be excited if there were to be a supernova in our Galaxy near enough for detailed study to be possible.

Theoretical models of supernovae

Given that the *observations* of supernovae are not at present capable of telling us how much mass they lose, we must turn to *theories* of supernova explosions to see whether they can help us. We should perhaps start by asking how it is that a supernova can suddenly increase in brightness by such a large factor. It cannot simply be that there is a sudden increase in the energy released in the interior of a star, which is carried to the surface by normal means. The additional energy would take a time of order of the thermal timescale of the star

$$t_{th} = GM_s^2/r_s L_s \qquad (7.1)$$

to do that and for most stars the thermal timescale is 10^4 years or more. In the case of the Sun, if there was a large increase in the rate of energy release in the centre, radiation would take more than 10^7 yr to carry the excess energy to the surface. In contrast, in a supernova, changes in brightness are occurring in a time which is comparable with the dynamical timescale

$$t_d = (r_s^3/GM_s)^{\frac{1}{2}}, \qquad (7.2)$$

which is measured in hours or days for most stars.

118

A deceptively simple suggestion is that an explosion in the interior of a star removes all of the outer layers of the star and exposes the hot interior. Thus, since the luminosity of a star is

$$L_s = \pi a c r_s^2 T_e^4, \qquad (7.3)$$

an increase in luminosity by a factor of 10^6 or 10^9 requires $r_s^2 T_e^4$ to increase by that factor. Even the removal of all the material at temperatures less than 10^7 K, with some reduction of the radius of the star, would increase the luminosity of an ordinary star to a supernova value for a short time. If the outside of a star could be removed easily, it would be relatively simple to produce a supernova model. One cannot, however, wish the mass away as easily as that. Even if it is moving outwards as a result of the explosion, it is still there to impede the escape of radiation from the interior and only strong departures from spherical symmetry in the explosion could make a model of this type at all plausible.

In fact, some calculations of violently exploding stars have suggested that essentially all of the energy goes into the motion of the ejected matter and that no visible supernova results. It seems likely that in *any* supernova explosion *much* of the energy is carried outwards from the site of the explosion in strong compression waves. As the waves travel outwards they steepen into *shock waves*, which then heat the outer layers to such an extent that thermonuclear reactions occur in them. Thus heavy elements may be synthesised during a supernova explosion as well as before it. If an outer layer, from which radiation can escape, is raised to a sufficiently high temperature a visible supernova results. It is nevertheless possible that the visible energy is a small fraction of the total supernova energy and that the total number of supernovae (defined as stars which explode and lose a substantial amount of mass) may exceed the number of visible supernovae. This describes *some* current theoretical ideas about the origin of the high luminosity of a supernova but there are still considerable uncertainties in this subject.

Cause of supernova explosion

We must now discuss the trigger mechanism, which sets off the explosion in the first instance. A large number of proposals have been made for the cause of supernova explosions. We are unable to discuss them all in detail here but we can make a few comments on two of them*. We can first remark that in some theories the explosion occurs at the centre of the star whilst in others it occurs some way out from the centre. If the explosion occurs at the stellar centre, there is a possibility that the star could be completely shattered

---

* The theories are not all competitive and different mechanisms probably cause explosions in stars of different masses.

by the explosion and that no substantial remnant remains. This is, of course, not inevitable and any mass loss from zero to the total mass can occur. If the explosion occurs in a shell around the centre, there will almost certainly be an imploded remnant and, the further out from the centre that the explosion occurs, the larger the remnant is likely to be. These possibilities are illustrated in fig. 49. We now give a brief description of theories of the two types.

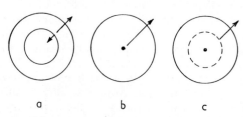

a   b   c

Fig. 49. Three possible types of supernova explosion: (a) Explosion in shell; outside ejected and inside imploded. (b) Explosion in centre and no remnant left. (c) Explosion in centre but only mass above dashed circle removed from star.

As a star evolves, the material in its interior initially behaves like a perfect gas with the equation of state

$$P = \mathscr{R}\rho T/\mu \qquad (7.4)$$

where P, ρ and T are the pressure, density and temperature of the material, μ is the mean molecular weight of the material (mean particle mass in terms of the mass, m_H, of the hydrogen atom) and $\mathscr{R}$* is the gas constant ($8 \cdot 26 \times 10^3$ J K^{-1} kg^{-1}). As the central regions of the star contract and increase in density, there is a possibility that they become an imperfect or *degenerate*† gas in which to a high degree of approximation the equation of state is of the form

$$P = f(\rho) \qquad (7.5)$$

so that the pressure is independent of temperature. If energy-releasing nuclear reactions start in a non-degenerate gas, the

* This gas constant $\mathscr{R}$ differs from the usual SI gas constant R by a factor of approximately 10^3. They are essentially equal in c.g.s. units, but differ in SI units because, while the unit of mass is different, the mass of a mole is not.

† The properties of a degenerate gas have been described in some detail in *The Stars*. The occurrence of degeneracy is caused by the operation of the Pauli exclusion principle. When matter is compressed electrons are forced by the exclusion principle into higher energy states than they would normally occupy. At sufficiently high densities, the states they occupy are determined by the density independent of the temperature. The electron pressure then depends on density alone and this pressure is very much greater than the ion pressure so that (7.5) results to a good approximation. As the temperature rises the critical density at which degeneracy occurs also rises.

temperature rises *locally*. Through eqn. (7.4) the pressure then rises and this causes an expansion and cooling. The star possesses a self-regulating mechanism. In a degenerate gas a rise in temperature is *not* followed immediately by a rise in pressure. As the energy release from nuclear reactions is highly temperature dependent, this leads to a further temperature rise. This process continues until the temperature becomes so high that the gas becomes perfect again. At this stage the pressure does begin to rise rapidly and expansion may occur too rapidly for the self-regulating mechanism mentioned earlier to have time to act. If the material is initially highly degenerate, such a high rate of release of nuclear energy may occur that it is sufficient to explode the star, in some cases so violently that no remnant is left.

Explosive carbon burning

In low mass stars ($\leqslant 1\ M_\odot$) the onset of helium burning occurs in very degenerate material. Although there is then a very high release of energy from helium burning, calculations do not at present suggest that such stars explode and become supernovae. In stars of somewhat higher mass, helium burning occurs in non-degenerate material, but after the helium is exhausted the central density rises and the central regions are highly degenerate before carbon burning starts. Carbon burning can then occur explosively and some calculations suggest that a star of about $5\ M_\odot$ could explode as a supernova and leave no significant fragment behind. It is very important that these calculations be checked and that their predictions be compared with observation, as this could be a very important source of heavy elements, if those elements heavier than carbon can be produced by thermonuclear reactions *during* the explosion. Calculations *have* been made of explosive carbon burning which suggest that substantial quantities of elements between carbon and iron can be produced and these calculations are mentioned further below on page 141.

Iron-helium phase change

A second type of supernova model may be applicable to more massive stars which reach the end point of normal thermonuclear evolution without their material becoming degenerate. After the central regions of a star have been converted into elements such as iron and nickel, they are unable to release any more energy by nuclear reactions. The regions continue to contract and to heat up and eventually the heavy nuclei dissociate into a mixture of helium and neutrons, by reactions such as

$$^{56}\text{Fe} \to 13\ ^4\text{He} + 4\text{n}, \tag{7.6}$$

as has already been described on page 90 in Chapter 5. Reaction (7.6) requires a large amount of energy which can only be provided by

121

further release of gravitational energy. Thus the star's central regions collapse even more rapidly. Material further out in the star is then brought to a high temperature very rapidly and its remaining nuclear fuel may be ignited explosively. As a result the outside of the star may be ejected, whilst the inside is imploded.

Although rough estimates of the amount of energy available in such an explosion suggest that most of the mass of the star *could* be ejected into space, detailed calculations have so far failed to show that this happens. The results are not conclusive because the calculation of a stellar explosion is a difficult task. To do this it is necessary to solve a complicated set of equations which give all of the physical quantities such as temperature, density and outward velocity as a function of both radial coordinate and time. It is complicated because some of the physical quantities vary by many orders of magnitude from the centre to the surface of the star and by the occurrence of shock waves in which an almost discontinuous change in a physical quantity can occur. Some of these difficulties are being overcome by the development of larger and faster computers. As far as the explosions of massive stars are concerned, there is at least the suggestion that they may not eject *most* of their mass.

Pulsars and supernovae

Until recently there was no observational information about the remnants left behind by supernova explosions. For a long time it was believed that they must be white dwarfs, but that was only because there was no belief in the possible existence of neutron stars and black holes. In fact, no white dwarf was shown to be associated with a supernova. In 1968 the discovery of a new type of astronomical object, the *pulsar*, was announced. Pulsars emit periodic bursts of radio waves with a very short period and the pulse of radiation may only last 30 ms in a period of 1 s. Such a well-defined pulse of radiation can only come from a very small object; if the emitting region were too large, the time of arrival of signals from different parts of the region would spread the pulse out in time as is indicated in fig. 50. At present it appears that only neutron stars are small enough to be the source of the pulsar emission and the most popular view is that

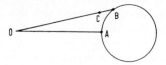

Fig. 50. Light travel time from an extended source. Light from two points A to B takes different times to reach an observer O. Even if a sharp pulse is emitted simultaneously over the source between A and B, it must be spread out on arrival by the time it takes light to travel from B to C.

pulsars are rotating neutron stars with radio emission coming from one or two regions near their surfaces.

The pulsar with shortest period (0·03 s) is in the Crab nebula and it is also the only pulsar which has been observed to emit pulses of visible radiation, X-rays and γ-rays. It is natural to suppose that it is the remnant of the 1054 supernova and that other pulsars are associated with earlier supernovae; it is, however, noticeable that no pulsar has been found in the region of the supernova of 1572 and 1604. It is found that periods of individual pulsars increase with time and that long period pulsars are weaker emitters of radiation than those of short period. Assuming that short period intense pulsars evolve into long period weak pulsars, it is possible to estimate that this evolution takes of order 10^7 years. Only pulsars in a limited region of the Galaxy can be observed but it is then possible to estimate how many there are in the complete Galaxy. The number obtained is of order 10^5 and this, combined with an *average* age of those observed of half the value given above, suggests that one pulsar is produced each 50 yr. This is comparable with the rate at which supernovae are believed to occur (see page 133), so that a reasonable fraction of supernovae may leave neutron star remnants with masses less than the maximum mass of a neutron star, which is believed to be about 2 $M_\odot$.

Mass loss from other stars

These include (i) novae, (ii) planetary nebulae, (iii) main sequence and pre-main-sequence stars, and (iv) red giants. The loss of mass in most cases is even less well understood than the loss of mass from supernovae.

(i) Novae are quite spectacular. They increase in luminosity by several orders of magnitude and observations of Doppler-shifted spectral lines show that some mass is moving outwards from them. Detailed study shows that the total loss of mass is very small, being a fraction of 1 per cent of the mass of the star. This immediately shows that novae cannot play an important role in nucleosynthesis unless stars become novae very many times in their life history. Another way of demonstrating that they cannot be very important is to calculate the total rate at which all novae feed mass into the inter-stellar medium. It is now believed that all novae are close binary stars and that most of the *apparent* loss of mass is merely an exchange of mass between the two components of the system. Such mass exchange can occur if the radius of one component increases as it evolves to such an extent that the outer layers of the star are more strongly attracted gravitationally by the other star than by the star to which they initially belong. Such mass exchange often occurs quietly but it appears that in some circumstances it can occur very rapidly and give rise to the nova phenomenon.

123

(ii) The planetary nebulae are more serious candidates as the site of nucleosynthesis *at the present time*. Planetary nebulae are stars which are surrounded by a sphere or spherical shell of gas, which has almost certainly been ejected from the star at a previous stage, as the gas is observed to be expanding away from the star. The total system of star and nebula typically has a mass of about 1 $M_\odot$ and the mass that has been ejected may be about $0 \cdot 2\ M_\odot$. The planetary nebulae are not believed to result from very highly evolved stars so that they are only likely to be an important source of elements such as helium and carbon. Because of their low mass, planetary nebulae can hardly have played an important role in nucleosynthesis in the early life of the Galaxy as they cannot have evolved rapidly enough.

(iii) There is some evidence that mass loss is occurring from stars, which are in the stage of pre-main-sequence evolution before the first nuclear reactions start in their interiors, and from some main sequence stars. As an example the Sun is losing mass into space at a rate of between $10^{-13}\ M_\odot$ and $10^{-14}\ M_\odot$ per year. At this rate the total mass lost in the Sun's lifetime in the *solar wind* is a minute fraction of its total mass. It is believed that the Sun's loss of mass is associated with its possession of a region near to its surface where convection is carrying energy outwards and it is further believed that more extensive *stellar winds* may exist in stars which have more violent convection than the Sun. This loss of mass from pre-main-sequence and main sequence stars is irrelevant to the subject of this book, as the mass lost has not been processed by any nuclear reactions, except inasmuch as it tells us that star formation is a less efficient process than we thought; this mass loss essentially reduces the mass of the star formed.

(iv) There is also evidence that some *red giants* are losing mass at quite a high rate (up to $10^{-6}\ M_\odot$ per year has been observed in a stage of evolution which may last 10^6 yr). These stars have had time to convert some hydrogen into helium so that it is possible that the lost mass is enriched in helium, provided that the star has been able to circulate the helium from the centre to the surface. These stars cannot be an important source of heavy elements at this stage in their evolution, but the *indirect* effect of this mass loss may be greater than at first appears. If it occurs after helium burning has started and a substantial fraction of the star is lost, the star will have lost most of its hydrogen rich exterior. If the star continues its evolution and subsequently becomes a supernova it is possible that a larger *fraction* of the mass ejected in such a supernova explosion is in the form of heavy elements, as there is now no hydrogen rich envelope to be ejected. Alternatively, a high rate of mass loss in the red giant phase could reduce the mass of the star to such an extent that it never becomes a supernova. Thus the uncertainties in our knowledge of red giant mass loss may lead to uncertainties in our estimates of the total rate of nucleosynthesis in supernovae.

It appears from the discussion given above that we have not suggested any alternative to the supernova explosion as a major source of heavy elements. It is perhaps easy to see why this should be so. Heavy elements must be ejected into the interstellar medium either as a result of an explosion or by the steady loss of the surface regions. The only stellar explosions which we observe are novae and supernovae; there may be production of heavy elements in *invisible supernovae* as mentioned on page 119, but it is difficult to see how these could be detected. If, in contrast, the metals enter the interstellar medium by the steady loss of surface material, the stars which are losing mass must have a very high surface abundance of metals and observations suggest that such stars are extremely rare. It appears that at present supernovae are almost certainly the major source of heavy elements but we must bear in mind that we cannot observe the stars which produced heavy elements in the early stages of the galactic history. We shall return to this point later.

White dwarfs

The second most common type of star, after the main sequence stars, is the white dwarf and it used to be believed that they gave the best estimate for the occurrence of mass loss. The maximum mass of a white dwarf, according to theory, is $1 \cdot 2 \, M_\odot$ and those whose masses are known are below that limit. Stars less massive than $1 \cdot 2 \, M_\odot$ can only just complete their life history in the estimated age of the Galaxy. At one time the best estimate of the galactic age was considerably lower than at present; then it was thought that such stars could certainly not have evolved in the galactic lifetime and that they must have passed most of their life history as more massive stars and have then lost mass to bring them below the white dwarf limit. Now the situation is less certain; although it is clear that the lower mass white dwarfs must have been formed from stars which have lost mass. What is uncertain is whether most white dwarfs have been produced from such stars which were originally only a little more massive (such as planetary nebulae) or whether they have resulted from much more massive stars.

7.3   The circulation of material in the Galaxy

It should be clear from the previous section that the subject of mass loss from stars is very uncertain but that supernovae are still the only seriously suggested source of the heavy elements. In this section we ask what happens to the mass that is ejected in the explosion of a supernova, when it interacts with the interstellar medium. It may first be noted that, if there were no interstellar matter, the mass expelled from many supernovae would leave the Galaxy as its speed exceeds the escape velocity from the Galaxy. Thus, in the solar neighbourhood, the escape velocity is of order 3×10^5 m s^{-1}, whereas many supernovae eject matter with a speed of 10^7 m s^{-1}. Thus, if there were no interstellar matter, most of the heavy elements ejected by supernovae would alter the chemical composition of the inter-galactic medium rather than the Galaxy itself.

The interstellar material is mainly concentrated in the disk of the Galaxy and it has a mean density of about 10^6 particles m^{-3}. The density is not uniform and much of the gas is in clouds with a density one or two orders of magnitude higher. These clouds contain typically a few hundred solar masses of gas and they move with speeds of about 10^4 m s^{-1} through a gas of lower density. Their motion in the gravitational field of the rest of the galactic mass gives the gas disk a thickness of about 300 parsecs ($\approx 10^{19}$ m). If we now suppose that a supernova ejects material into interstellar space, how long is it before that material is indistinguishable from ordinary interstellar matter?

The Snowplough mechanism

We can make a rather crude estimate as follows. A typical supernova ejects material with a speed of order 10^7 m s^{-1}, which is about 10^3 times the random velocity of the interstellar gas. If this material acts in the same manner as a snowplough and gathers up all of the interstellar material ahead of it, it is possible to estimate how it is slowed down. If the ejected mass is m and its initial speed is V and at a later time the total mass of swept up gas and ejected gas is M and the speed is v, conservation of momentum implies that

$$mV = Mv \qquad (7.7)$$

In this estimate we are neglecting the initial momentum of the swept up gas, which we can do so long as v is very much greater than a typical interstellar gas velocity. As long as this is true, interstellar gas will continue to be swept up, but once v falls to about 10^4 m s^{-1} the regular motion will be broken up by collisions with interstellar clouds and the matter will then become part of the ordinary inter-stellar gas. At this time $M \approx 10^3 \, m$. Clearly this discussion is highly simplified because nothing like spherical symmetry would be maintained for as long as we have supposed, but it should give an answer which is of the correct order of magnitude.

This calculation suggests that the heavy elements ejected in any supernova explosion will *pollute* about a thousand times its own mass of interstellar gas. This means that any one object is unlikely to cause a significant change in the chemical composition of the material which it affects. We shall require very special circumstances if there is to be a sudden increase in the heavy element content of any region in the Galaxy. If the objects producing the heavy elements are reasonably well distributed through the Galaxy, we can expect that the resulting distribution of heavy elements will be smooth but not necessarily uniform. It appears that the material in a gas cloud in the Galaxy today which may contain 3 per cent of its mass in the form of heavy elements must have been affected by up to 100 supernovae, if the mass ejected by each supernova contains hydrogen and helium as well as heavy elements. This indicates that the similarity in the mix of heavy elements from star to star, which we have discussed on page 50 need not mean that all of the heavy elements have been produced in one process; instead all interstellar clouds out of which stars are formed may have been exposed to a representative sample of different types of supernovae.

We have already mentioned above that it is vital that there is some interstellar gas to be swept up by the material ejected by a supernova so as to confine it to the Galaxy. We should now check up that the thickness of the galactic disk is great enough to contain the material ejected by a typical supernova. Suppose $1\ M_\odot$ of material is ejected. It will be mixed into the interstellar medium when it has swept up $10^3\ M_\odot$. At a mean particle density of $10^6\ m^{-3}$, a sphere containing $10^3\ M_\odot$ has a radius of about 20 parsec so that it is comfortably contained within the galactic disk unless the supernova is very close to the edge of the disk. Of course, the material has to sweep up much less matter to reduce its velocity to below the escape velocity from the Galaxy; if material encounters very little interstellar matter compared to the average quantity, it could become a high velocity cloud which would subsequently fall back into the galactic disk. If each supernova ejected only $1\ M_\odot$, about 10^7 supernovae would be needed to affect all of the galactic gas assuming that it represents about 5 per cent of the galactic mass. We shall have more to say about such numbers shortly.

7.4 *Statistics of element production*

Here we come to what is perhaps the weakest link in our chain of argument. It is particularly weak because now we are bringing together observations and theories which both contain considerable uncertainties and are hoping to decide whether ordinary stars can have been responsible for all of the nucleosynthesis that has occurred in the Galaxy since its formation. The information which we have at our disposal can be summarized as follows:

(i) The oldest stars in the halo of the Galaxy contain a very small fractional abundance of elements heavier than helium.

(ii) The stars in the disk of the Galaxy have a heavy element content which does not differ greatly from that of the Sun, and this is true of stars which are only slightly younger than the old halo stars.

(iii) Most stars for which a helium content can be determined have a substantial helium content of order 25 per cent by mass but it is *possible* that the oldest stars contain essentially no helium.

From (i) and (ii) we deduce that:

(iv) The rate of production of heavy elements must have been much higher in about the first 10 per cent of the Galactic life than it has been since.

From (iii) we see that:

(v) It is necessary to investigate whether essentially all of the helium in the Galaxy could have been produced in its life history in case the helium is not primaeval.

From the discussion earlier in this chapter we deduce that:

(vi) Of the stars that we observe today, only supernovae seem capable of producing heavy elements at the required rate.

(vii) If the heavy elements have been produced in supernovae, the interstellar medium should have a heavy element content which varies rather smoothly from place to place.

Observations indicate that

(viii) The chemical composition of the interstellar medium is not uniform and and in particular there is a suggestion that the heavy element content is higher near the galactic centre than elsewhere. However, there is no clear indication of large variations in chemical composition in small regions of space.

In discussing the history of nucleosynthesis in the Galaxy, it must be remembered that:

(ix) The objects which were responsible for the early very high rate of nucleosynthesis may be of a type which is not common today.

(x) The total rate of production of heavy elements must have exceeded the useful rate of production, because allowance must be made for the heavy elements locked up in the remnants of exploded stars.

The total mass of the Galaxy and the proportion in the forms of hydrogen, helium and heavy elements are all very uncertain but to a first approximation we can say:

(xi) The total useful production of heavy elements must have been of order $2 \times 10^9 \, M_\odot$.

(xii) If the helium was not primaeval, the useful production of helium must have been about ten times the useful production of heavy elements, since young objects today have about ten times as much helium as heavy elements.

Helium production in stars

We first discuss the rate of helium production in stars. In the preceding chapter we have discussed the possibility that essentially all of the helium in the Galaxy was present before any of the stars were formed. We have pointed out the attractions of that theory but we have also indicated that we cannot at present regard the observations in its favour as being completely secure. It is therefore desirable to discuss the alternative possibility that all of the helium has been produced *since* the Galaxy was formed.

As helium is the first element produced by nuclear reactions in stars, it might be thought that it could readily be accounted for by stellar nucleosynthesis in the galactic lifetime. As soon as we examine the problem in detail, difficulties arise on account of both the *total amount* of helium required and the *relative amount* of helium and heavy elements. In the first place we can repeat that the observations indicate that much of the nucleosynthesis must have occurred rather early in the galactic lifetime and this implies that, if the helium and heavy elements have been produced in stars, they must have been produced in rather massive stars because only such stars can have evolved in the time available. As we have already described, massive stars partake of a whole sequence of nuclear fusion reactions whereby, after hydrogen has been converted into helium, helium is converted into carbon and heavier elements. As has been shown in fig. 45 on page 113, the amount of unburnt helium in a star is at any time rather small and, after the central helium has been converted into carbon, it is less than the amount of carbon and heavier elements. If the amount of helium to be released into the interstellar medium is to be ten times greater than the amount of heavy elements, a rather carefully contrived type of mass loss must occur; instability must eject the hydrogen and helium layers but very little of the heavier elements and no further nucleosynthesis must occur in the ejected material during the explosion.

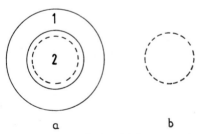

Fig. 51. A massive star before explosion (a) has an outer region (1) which is mainly H and He and an inner region (2) which is entirely heavy elements. If a substantial amount of heavy elements is not to be ejected by the explosion, the mass below the dashed curve must form a massive remnant (b).

Even if mass loss of that type does occur some problems remain. In the first place the explosion is likely to leave a fragment which is comparable in mass with the ejected material (see fig. 51). This fragment will almost certainly be too massive to become a white dwarf or a neutron star and it cannot become unstable again without upsetting the carefully contrived balance between the quantity of helium and heavy elements. If the balance is to be maintained, it is necessary for it to become a collapsed object. If we accept the possibility that all of the helium has been produced in stars, we may need to believe that a significant fraction of the mass of the Galaxy is locked up in collapsed objects, black holes, which are unobservable except possibly for their gravitational attraction on nearby stars. Although it may not be attractive to assume that much of the galactic mass is in an unobserved form, our present knowledge of the total mass of the Galaxy and of the parts that we can observe does *not* rule out this possibility.

The past luminosity of the Galaxy

There are also difficulties concerned with the total production of helium. If even 10 per cent* of the mass of the Galaxy has been converted into useful helium, it seems that at least twice this amount must have been initially converted to allow for subsequent conversion of helium to heavy elements and for the helium which is still inside stars of low mass, which have not yet completed their life history. If 20 per cent of the galactic mass has been converted from hydrogen into helium in the galactic lifetime, the total energy release from this process will be an underestimate of the total integrated luminosity of the Galaxy. Thus, assuming a galactic mass of $1 \cdot 5 \times 10^{11} M_\odot$ and an age of 10^{10} yr, the total energy release is of order 4×10^{55} J and the average luminosity is more than 10^{38} W. As a large fraction of the nucleosynthesis must have occurred in the first 10^9 yr the peak luminosity must have been at least 10^{39} W; this compares with around 2×10^{37} W at present. This would mean that the Galaxy once had a luminosity comparable with that of the most luminous objects (quasars) of which we have knowledge.

The high predicted luminosity of the Galaxy in its early phases is of interest in suggesting that we may not be living in the liveliest phase of galactic history, but it is no clear argument against helium production in stars. A greater difficulty is that we are demanding that at least 10 per cent of the mass of the Galaxy be converted into useful helium, while on page 113 we have stated that it is unlikely that as much as 10 per cent of the mass of any individual massive star can be converted into useful helium. This implies that, if all the helium has been produced in stars, essentially the whole galactic mass must

---

* This is a low mean value between 25 per cent in young stars and a possibly zero content in the oldest.

have been converted into massive stars when the Galaxy first formed. Even then we probably require more than one generation of such massive stars so as to produce all of the helium; the mass expelled by the first generation of massive stars may have condensed into a further generation of massive stars and some further conversion of hydrogen to helium can have occurred. The problem with this process is that in each generation of stars a large amount of matter ends up in the form of collapsed objects and we are led to the view that they may be the principal component of the Galaxy. Even this cannot be regarded as conclusive evidence that the helium was *not* produced in stars, but it would be easier to understand the present helium abundance if most of it were primaeval; as mentioned earlier it is vitally important to know whether any stars exist which contain virtually no helium.

Helium production in very massive stars

Until recently it was generally believed that there was no way around the difficulties described above, but there is now a suggestion that the helium could have been produced in *very* massive stars. What is really needed, if helium is to be produced usefully in stars, is that the region of the star that is affected by the conversion of hydrogen into helium should be as large as possible and that the star should become unstable after hydrogen has been converted into helium and before much helium has been converted into carbon and heavy elements. In all massive stars the initial hydrogen burning affects a large fraction of the mass; hydrogen burns in a core which is kept well mixed by convection currents and the initial mass of the core increases with stellar mass as is shown in fig. 52. Conventional massive stars do not,

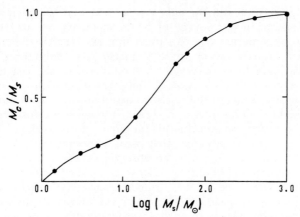

Fig. 52. The mass in a stellar convective core (M_c) is shown as a function of stellar mass. These theoretical results are from the work of several authors and the true curve may be somewhat smoother.

however, become unstable until fairly late in their evolution and that gives rise to the difficulties in relative production of helium and heavy elements, which have already been discussed.

The class of star which may meet the requirements is more than about 50 $M_\odot$ or 100 $M_\odot$ in mass. Such stars are believed to be unstable to radial oscillations of increasing amplitude; in any star small disturbances from equilibrium will continually arise spontaneously but, whereas in most stars all disturbances will be damped rapidly, in these stars this particular type of disturbance grows. Calculations of the initial growth of the instability suggest that the star might be disrupted long before it has completed its main sequence life and it is generally supposed that there is a maximum mass for main sequence stars because more massive stars are disrupted rapidly. Very recently this conclusion has been challenged.

It is relatively easy to show that a system is unstable to small disturbances but it is much more difficult to follow the growth of the instability to a large amplitude. Recent attempts to study these very massive stars have suggested that they may lose mass more slowly than was previously supposed and that it is possible that they lose most of their mass after much of the hydrogen has been converted into helium but before the helium has been converted into heavier elements. If this is true and if when stars first formed in the galaxy they were preferentially of mass 100 $M_\odot$ and above, such massive stars may have been a major source of galactic helium, but at present there are far too many uncertainties in the theory to place any reliance on it.

The production of heavy elements
the frequency of supernovae:

As the supernova is still the only type of star that has been seriously proposed as the major source of heavy elements, we will base our discussion on supernovae. We must first ask what is the frequency with which supernovae occur in a galaxy. As only three or four supernovae have been observed in our Galaxy in the last thousand years, it *appears* that the Galaxy only has a supernova every few hundred years, although the statistics based on such small numbers is very unreliable. This total is, however, unreliable for another reason. There is so much gas and dust in the *plane* of our Galaxy, in which most of the stars are situated, that many supernovae could occur in the Galaxy and not be seen by an observer on the Earth. The light from the supernova would be almost completely absorbed by the interstellar matter; see fig. 53. Supernovae are so bright that they can be observed in galaxies other than our own and, unless we are observing another galaxy in such an unfavourable alignment that we look straight through its disk of gas and dust, we shall not be prevented from seeing its supernovae by obscuration. This means, rather surprisingly, that it is easier to make an estimate of the frequency of occurrence of supernovae by observing other galaxies. A regular supernova search has been mounted by the Mount Palomar Observatory* for some years and many have been discovered. Various analyses have been made of the numbers which have been observed and it has been suggested that a reasonable estimate of the present frequency of supernovae in a galaxy like our own should be about 1 per 30 years; about 10 times as many supernovae as have been observed may have exploded in the Galaxy in the past thousand years.

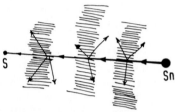

Fig. 53.   Absorption of light from a supernova in the Galaxy.   The light from a supernova, Sn, is considerably reduced by absorption and scattering by interstellar matter before reaching the Sun, S.

This is, of course, an estimate of the present rate of supernova explosions in our own Galaxy and the number may have been very different in the early life of the Galaxy. If we suppose as a working hypothesis that, after the initial burst of star formation and element production, the Galaxy has had a relatively steady rate of supernovae, there should have been about 3×10^8 supernovae in the past 10^{10} years. If half of the heavy elements which we observe today have been produced during that time and if they have all been produced in supernovae, we require about $3\ M_\odot$ of heavy elements to be produced in each supernova. Bearing in mind that some hydrogen and helium is likely to be expelled in any supernova explosion, the total mass ejected may need to be $5\ M_\odot$ or more. Although these numbers are very uncertain, there is no *direct* evidence that supernovae are at present ejecting such large masses of matter and many of the theoretical models of supernovae give a smaller loss of mass. It is true that the model of the $5\ M_\odot$ supernova which has been described on page 121, which is disrupted completely, might give the correct amount of heavy elements, but we still have the problem that we do not yet have conclusive observations of such large rates of mass loss.

All of the numbers that have been used in the above discussion are very uncertain so that although there is a discrepancy in these figures, it is certainly too early to say that supernovae are not the main source of heavy elements. In addition, as has been mentioned earlier in this

* Now part of the Hale Observatories.

133

chapter, many theoretical studies of supernovae have predicted the existence of stars which explode and lose mass without ever having a very high luminosity. It is possible that the *visible* supernovae are only a small fraction of all the stars which end their life history in an explosion and expel heavy elements into interstellar space. Clearly, if the total number of exploding stars were 5 or 10 times the number of visible supernovae, the mass of heavy elements to be produced in each explosion would be reduced to more manageable proportions. To discuss this possibility we need an independent estimate of the number of stars dying each year and we will discuss this briefly below, after we have made some further comments on the connection of supernovae with cosmic rays.

The energy input into cosmic rays

We have already suggested several times that cosmic rays may be produced in supernova explosions and it is now possible to ask how much energy must be put into cosmic rays in a typical explosion, if this is true. Direct observations of cosmic rays can only be made in the Earth and its neighbourhood, but, since the cosmic rays reach the Earth isotropically, it is believed that the density of cosmic rays near the Earth is typical of a much larger region in the Galaxy. If it is supposed that the cosmic rays are mainly in the disk of the Galaxy and that the cosmic ray energy density throughout the disk is about $1\cdot6 \times 10^{-13}$ J m^{-3} as it is near the Earth, the total energy in the form of cosmic rays is about $1\cdot3 \times 10^{48}$ J, using the dimensions of the galactic disk given in Chapter 1. In Chapter 4 we have seen that the cosmic rays stay in the Galaxy for about $1\cdot2 \times 10^{14}$ s, which means that there needs to be a continuous input of energy into cosmic rays at an average rate of about 10^{34} J s^{-1}. If this energy is produced in supernovae which occur at a rate of one supernova every 10^9 s, we need 10^{43} J of cosmic ray energy per supernova. Estimates of the total amount of energy involved in supernova explosions, obtained from both observations and theories, tend to lie in the range $10^{42} - 10^{45}$ J, with the observational figures lying in the lower part of the range.

Comparison of these figures suggests that it is not impossible that all cosmic rays are produced in the explosions of supernovae but that, as mentioned earlier, quite a large fraction of the energy of the explosion may need to be given to a very small fraction of the ejected material. Once again the problem would be eased if some fraction of the cosmic rays were produced in explosions which are not seen as visible supernovae.

The distribution of stellar masses

We now ask whether we have any good idea of the rate at which stars, which could become supernovae, visible or invisible, may have been produced during the life history of the Galaxy. To do this we must study the number of stars in different mass ranges to try to discover what fraction of the galactic mass is in the form of potential super-novae.   Counts of stars in the solar neighbourhood, which is the only region of the Galaxy which can be studied in great detail suggest that if $F(M)\mathrm{d}M$ is the number of stars with masses between M and $M+\mathrm{d}M$ formed out of a given mass of gas

$$F(M)\propto M^{-7/3}. \tag{7.8}$$

Although this law is certainly not exact, it appears to be a reasonable approximation in the mass range in which stars can be studied easily but the form is much more uncertain at the low mass end, where stars are too faint to be observed easily.   This is unfortunate because much of the mass of the Galaxy is concentrated in low mass stars. The function $F(M)$ is called the *initial mass function*.

Different authors have suggested that the function $F(M)$ should be cut off at a lower mass of about $0 \cdot 1\ M_\odot$ or $0 \cdot 025\ M_\odot$.   As very little of the total mass is in massive stars, if a law of the form (7.8) holds, the upper mass limit is not very important.   We take it to be $100\ M_\odot$. If we now recognise that the important mass range for stars which can produce significant amounts of heavy elements is for masses greater than about $5\ M_\odot$, we can calculate what fraction of a given mass of gas goes into stars of such masses.   For the two values of the minimum mass we obtain

$$\left.\begin{array}{r} \text{Mass } (M>5\ M_\odot)/\text{Total Mass} \approx 0 \cdot 2 \\ \approx 0 \cdot 1. \end{array}\right\} \tag{7.9}$$

We can in addition ask what fraction of the mass would be in the form of stars less massive than $1\ M_\odot$; whenever these stars were formed in the life history of the Galaxy, they will not have had time to complete their life history, if present estimates of the galactic age are correct.   For the two values of the minimum mass we obtain

$$\left.\begin{array}{r} \text{Mass } (M<M_\odot)/\text{Total Mass} \approx 0 \cdot 60 \\ \approx 0 \cdot 75. \end{array}\right\} \tag{7.10}$$

The past rate of star formation

Can *this* initial mass function have been valid throughout the whole of galactic history?   If it has, it seems possible that all of the heavy elements which we observe today could have been produced in stars of mass greater than $5\ M_\odot$.   From (7.9) they would only need to convert about $1/5$ of their mass into useful production of heavy elements, even if there has only been one generation of stars, given the present estimate of $2\times10^9\ M_\odot$ of heavy elements in a galactic mass of between 1 and $2\times10^{11}\ M_\odot$.   In addition, since a star of $5\ M_\odot$

completes its life history in about 10^8 years, the need for about half of the heavy elements to have been produced very early in galactic history could be met if half or more of the galactic gas were involved in the initial burst of star formation. This looks very satisfactory until we consider the implication of the numbers given in (7.10). In such an initial burst of star formation, more than half of the mass converted into stars would have gone into stars less massive than 1 $M_\odot$. Such stars would not have completed their life history and should still be observable today with very low abundances of heavy elements. Such a large number of low mass stars with very low metal abundance is not observed and we are forced to the conclusion that the initial mass function (7.8) cannot have been valid throughout galactic history and that the initial generation of stars must have been deficient in stars of low mass compared to stars formed more recently.

This result is of interest because we have already seen on page 131 that, if all of the helium in the Galaxy has been formed in stars, there must have been a preponderance of massive stars in the first generation of stars formed. From the discussion which we have given of the helium abundance, it seems rather difficult for the helium and the metals, in the correct relative proportions, to have been produced in the *same* massive stars, but it is not impossible that the helium and the metals could both have been produced in a succession of early generations of massive stars. In this case the helium may have been produced in extremely massive stars and the metals produced in a subsequent generation of rather less massive stars. If this explanation is correct, as mentioned on page 131, a significant fraction of the galactic mass may now be in the form of black holes.

If we suppose that, after the initial production of massive stars, the mass that was expelled from these stars together with the gas that had not yet condensed into stars produced stars with a mass function comparable with (7.8) at an approximately constant rate, we obtain a crude estimate of the number of stars with masses greater than 5 $M_\odot$ which could have exploded in the last 10^{10} yr. If 50 per cent of the known galactic mass was involved in these later generations of stars and if in accordance with (7.9) above 10 per cent of this mass went into stars of mass greater than 5 $M_\odot$, about $1 \cdot 5 \times 10^9$ stars could have been formed. Over a period of 10^{10} yr, one star should have exploded every 6 or 7 years. These estimates are so crude that there is certainly no clear discrepancy with the observed supernova rate but there is certainly a possibility that the invisible supernovae mentioned earlier may also be important.

Models of galactic nucleosynthesis

With the aid of a computer it is possible to refine the very crude arguments that we have given above and to produce a model of the chemical composition of the Galaxy as a function of time. To do this two types of assumption must be made. These are:

(i) The form of the initial mass function. This is taken to be a function not only of the stellar mass but also of time since the formation of the Galaxy; such a form allows massive stars to be formed preferentially early in the galactic history. The dependence on time should be more realistically replaced by a dependence on density, temperature and chemical composition of the galactic gas, because ultimately it must be these quantities which determine the masses of the stars which are formed.

(ii) The element production in and ultimate state of stars of different masses and the time they take to complete their life history. The latter is quite well known but, as we have discussed earlier in this chapter, there is still considerable uncertainty in the mass lost by stars and its chemical composition.

If assumptions of type (i) and (ii) are made and if in the first instance all spatial variations in the Galaxy are neglected, it is possible to follow the time variation of the properties of the Galaxy to predict such quantities as:

(iii) The present composition of the interstellar medium.

(iv) The mass fractions of the Galaxy which should at present be in the forms of gas, various types of stars and black holes.

(v) The proportions of stars with different metal contents relative to hydrogen.

The quantities involved in (i) to (v) are all known to some approximation and the crucial question is whether, given all of their uncertainties, it is possible to assume that all of the heavy elements have been produced in stars in the galactic lifetime. Several groups of workers have recently produced models of galactic nucleosynthesis of the type described. Their assumptions and their conclusions differ and the best that can be said at present is that it does not seem impossible that the heavy elements have been produced at the required rate in ordinary stars. Future models of galactic nucleosynthesis will have to be more detailed; account will have to be taken of the spatial structure of the Galaxy and of consequent variations of chemical composition from place to place within the Galaxy.

Nucleosynthesis in massive objects

Although there is no obvious reason to discard the idea that almost all nucleosynthesis has occurred in ordinary stars, an alternative proposal has been made which is that a substantial amount of nucleosynthesis has occurred in objects much more massive than ordinary stars. We now discuss this idea and the reasons why it has been suggested.

137

We have mentioned on page 50 that there is an indication from observations that the heavy element content is higher near the centre of the Galaxy than in other regions. This may simply mean that the rate of supernova explosions per unit mass of material has been higher near the galactic centre than elsewhere, but it could also mean that explosions on a much larger scale have occurred near the galactic centre and have produced heavy elements. There is also the evidence mentioned on page 15 for very violent explosions in the *centres* of *some* galaxies which seem to be releasing as much energy as 10^{10} supernovae, and it seems unlikely that such a violent explosion can have failed to involve some nucleosynthesis. Could such explosions, or rather less violent ones, have played an important role in *galactic* nucleosynthesis?

Although violent galactic explosions are observed to occur there is at present no satisfactory theory of how they occur. The energy involved in the explosions is so large that nuclear energy alone seems unable to account for them. It is generally supposed that the gravitational energy released by a massive body, collapsing so that the gravitational energy release becomes comparable with its rest mass energy or

$$2GM_s/r_s c^2 \approx 1, \qquad (7.11)$$

may be responsible for the explosion. According to the general theory of relativity the gravitational energy release can certainly never exceed the rest mass energy and in practice probably does not exceed 10 per cent of the rest mass energy. This is, however, still comfortably in excess of the maximum nuclear energy release, which is just under 1 per cent of the rest mass energy.

Nucleosynthesis in little bangs

As there are no well defined models of galactic explosions, it is difficult to calculate how much nucleosynthesis occurs in them but some attempts have been made. Two distinct approaches can be identified. The first admits the possibility that we know so little about the origin of the Universe that the explosions may represent bursts of *creation*. On this view we have neither a *big-bang* theory of creation nor a *steady state* theory but creation in a sequence of *little bangs*. The second approach assumes that the galactic explosions must have occurred in the framework of ordinary astronomy, as opposed to new creation whenever it is needed, so that the large masses must have first formed in the centres of galaxies by condensation out of the general interstellar gas and then have exploded.

In the case of the little bang it is assumed that there is a pocket of creation in the centre of a galaxy. Matter is created in conditions of very high temperature and density and expands outwards at a velocity that can be chosen arbitrarily, as the mechanism of the creation process is completely unknown. In the early stages of such an expansion, a little bang is not very much different from a big bang, as discussed in the previous chapter, except that the density in the early stages can be taken to be much higher, at a given temperature, as it is no longer constrained by the *present* temperature and density of the Universe. We can expect such a little bang to produce substantial quantities of helium and, since higher densities encourage higher helium production, more helium can be obtained per unit mass than in the ordinary big-bang theory. If the big-bang origin of helium should be refuted, the little-bang theory is at least a plausible alternative.

Higher densities also encourage further nuclear reactions producing elements heavier than helium, but it is very difficult to produce substantial quantities of these elements, unless the parameters of the little bang, such as the maximum temperatures and densities attained and the expansion speed following creation, are chosen very carefully indeed. Since the production of heavy elements by a little bang is not impossible, such a theory might be considered, if future observations indicate that the heavy element content of the Galaxy is substantially higher in the central regions than elsewhere. There is, however, an outstanding difficulty in that little bangs do not seem to produce the *mix* of heavy elements observed in the solar neighbourhood; this mix would also need to be different in regions in which little bangs were important. Some results of little bang calculations are shown in fig. 54.

The main difference between the massive object produced in the galactic lifetime and the little bang is that the parameters of the massive object must be determined by its formation process. The nuclear reactions which occur in a massive object may be similar to those of one of the little bangs, but we can no longer choose which one arbitrarily. According to presently accepted laws of physics, a spherically symmetrical massive object will collapse under its self-gravitation to a state of infinite density and a nuclear explosion of *all* of its mass will be unable to halt that collapse. Unless the force of gravity becomes repulsive at very high densities, so that implosion can be followed by explosion, a spherical massive object will not be the origin of a large explosion in a galactic centre. If a massive object is rotating, it will flatten towards a disk and it may be possible for it to explode when it is in a highly flattened state. If this is possible and if the state immediately before explosion is one of very high temperature, nucleosynthesis along the lines of a little bang can occur. It would not be surprising for massive objects to form in the centres of galaxies. We can expect any gas which has not condensed into stars to have a tendency to fall to the region of lowest gravitational potential which is at the galactic centre, so that a massive object would be formed there.

139

There is one comment which applies particularly to nucleosynthesis in massive objects which has already been mentioned in connection with supernovae. Any matter ejected from a galactic explosion is likely to have such a high velocity that it has an escape velocity from the Galaxy. It can only be held within the Galaxy if it collides with sufficient interstellar gas for its velocity to be reduced below the escape velocity. In the case of the intense extragalactic radio sources, the explosion has apparently resulted in escape from the galaxy concerned. If useful nucleosynthesis is to occur in little bangs, it is necessary that they are not so large that all of the synthesized material escapes.

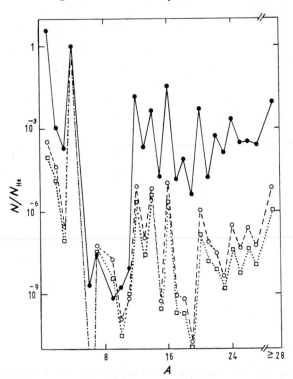

Fig. 54. Nucleosynthesis in little bangs. The results of two calculations (open circles and squares) are compared with solar system abundances. Although the relative abundances of moderately heavy elements can be reasonably reproduced the total quantity of heavy elements relative to helium is far too low.

Production of specific elements

All of the detailed discussion of the chapter has been concerned with how the total amount of the heavy elements and helium may have been produced during galactic history. No attempt has been made to account for the observed abundances of *individual* elements. To some extent this procedure is justified; as we have mentioned on page 50 in Chapter 3, the *mix* of heavy elements is relatively constant from star to star and it thus makes sense to treat them as a single entity. This does not mean that it is not very important to discuss how the observed abundance of each element has been produced and a considerable amount of study has been devoted to this problem. The reason why we have not discussed the fine details of either the observations or the theories is one of length and complexity. If we were to discuss the observations and production of each individual element, this book would need to be much longer and more complicated than is appropriate to the present series. We will, however, end this chapter by making a few brief remarks about the production of a few selected elements and isotopes.

Explosive nucleosynthesis

In our discussion of element production in supernovae we have mentioned that most of the heavy elements could be present in a supernova before it explodes, if they have been produced by fusion reactions in earlier stages of the star's evolution, or that the elements could mainly be synthesized in explosive nuclear reactions at the time of the supernova event. In the past the former view was favoured but recent calculations of explosive nuclear reactions have strongly

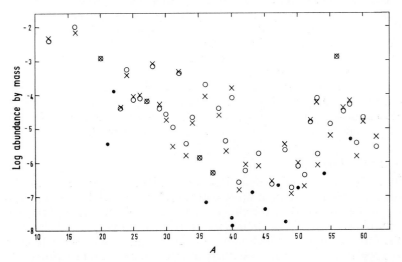

Fig. 55. Explosive nucleosynthesis. Abundances produced by explosive burning of carbon, oxygen and silicon (crosses) are compared with solar system abundances (open circles). The solid circles are observed abundances of nuclei which are not produced in substantial amounts by explosive nuclear reactions.

141

suggested that most nucleosynthesis might occur during the explosion. Explosive carbon burning could occur in the supernova models which we have mentioned earlier and could produce substantial amounts of heavy elements. Even more promising is explosive nucleosynthesis in more massive ($\sim 30 \, M_\odot$) stars whose chemical composition before the explosion would be that of the type shown in fig. 46 on page 114. In such a star explosive carbon, oxygen and silicon burning could occur in different regions and the relative abundances of elements between carbon and iron predicted by a recent calculation are shown in fig. 55. The relative abundances are compared with those observed in the solar system material. Although there is not complete agreement between the observations and the calculation, there is a remarkable similarity and this encourages the view that most of the isotopes between carbon and iron could have been produced by explosive nuclear reactions in massive supernovae.

The abundance of nitrogen

One isotope which is not produced very easily is ^{14}N. If we suppose that the Galaxy originally contained only hydrogen and helium, hydrogen must have burnt through the PP chain in the first generation of stars. This would have been followed by helium burning which produces no ^{14}N and by carbon burning, explosive or otherwise, which also produces no nitrogen. The next generation of stars would contain carbon and oxygen but no nitrogen, but some of them would now burn hydrogen through the CN cycle (see details on page 84.) In the CN cycle nitrogen is produced and, although subsequently it is processed back into carbon, nitrogen is much more abundant than carbon when the reaction is proceeding steadily. This is believed to be the main source of ^{14}N. As ^{14}N cannot be produced immediately nucleosynthesis starts, it appears that the oldest stars which we can observe should be deficient in nitrogen compared to younger stars. Recent observations have suggested that this is true and, if confirmed, they would be a very good confirmation of nucleosynthesis theories.

The r- and s-processes

The elements heavier than iron are believed to be produced by neutron capture reactions on elements such as iron itself. As the iron has to be produced before the process is possible, it seems likely that very old stars should be even more deficient in the very heavy elements than in iron and its neighbours. Observations of some very old stars suggest that this is true. As there are two neutron capture processes, the r-process and the s-process, we can ask whether we should expect isotopes produced by each process to appear equally early in galactic history. All of the early theoretical predictions suggested that the first r-process isotopes should have been formed earlier in galactic history

142

than the first s-process isotopes, so that s-process isotopes should be absent from the oldest stars. This prediction followed from ideas about when neutrons for the two processes could become available.

The r-process is believed to take place in explosive events such as a supernova explosion; amongst the reactions which could provide neutrons in such an explosion are the iron-helium phase change which has been described on page 90. Supernovae probably occur in all generations of stars and the r-process can then take place as soon as there are sufficient nuclei of intermediate mass to capture the neutrons. In contrast, for the s-process to occur, it appears that neutrons must be supplied in a major stage of stellar evolution. Particular proposals that have been made include the following. If a star has exhausted hydrogen in its interior and helium is burning there and then some instability causes hydrogen from the outside to be mixed into the regions containing helium and carbon, the following reactions, which produce neutrons, can occur.

$$\left.\begin{array}{l} p + {}^{12}C \rightarrow {}^{13}N + \gamma. \\ {}^{13}N \rightarrow {}^{13}C + e^+ + \nu, \\ {}^{13}C + {}^4He \rightarrow {}^{16}O + n. \end{array}\right\} \tag{7.12}$$

It has been predicted that these reactions would occur in stars of rather low mass which could not evolve in the very early stages of the Galaxy.

In fact, the s-process elements do *not* appear to be exceptionally underabundant in very old stars. This means that reactions such as (7.12) and the others that have been proposed must be capable of supplying neutrons for the s-process in stars massive enough to evolve rapidly or that an additional source of neutrons is required.

The production of lithium, beryllium and boron

It is now believed, as has been mentioned in Chapter 5, that these are produced by spallation nuclear reactions breaking up heavier nuclei. It used to be thought that these reactions occurred in the atmospheres of stars but it now seems probable the cosmic ray protons break up nuclei of such atoms as carbon, nitrogen and oxygen in the interstellar gas to produce much of the lithium, beryllium and boron which is observed. In addition some of these light elements *are* probably produced on the surfaces of those stars which possess electromagnetic fields which accelerate protons to high energies.

Summary of Chapter 7

This chapter has mainly been concerned with an attempt to obtain an estimate of the production of elements heavier than hydrogen in stars and of the way in which this has influenced the chemical compositions of the interstellar gas and of stars which have subsequently formed.

A discussion of the influence of nuclear reactions in stars on the observed chemical composition of stars and the interstellar gas is a complicated procedure. It is necessary to investigate:

(*a*) the amount of nucleosynthesis which occurs in stars of different masses;

(*b*) the manner in which stars expel matter into the interstellar medium;

(*c*) the mixing of this matter from stars into the general interstellar medium;

(*d*) the rate of formation of stars of different masses and the numbers of stars of different types that there are.

All of these steps are uncertain. In particular, if we wish to discuss the changes of chemical composition in the Galaxy throughout its life history, we must be interested in nucleosynthesis in stars formed early in galactic history and the stars which were most common then may well be different from the stars which are present today. The best conclusion that can be drawn from these investigations at their present stage is that *it does not seem improbable that all of the heavy elements have been produced in stars during the galactic lifetime but that it seems difficult to believe that the same is true of the helium.* It is clear from the observations of chemical composition of stars of different ages that nucleosynthesis must have occurred at a much more rapid rate in early galactic history than now and this in turn implies that the earliest generations of stars in the Galaxy must have been predominantly massive stars.

If the heavy elements are produced in stars, it appears that the main way by which they are distributed into space is through the explosions of supernovae; a major recent development is that it is now believed that nuclear reactions which occur during the supernova explosion are more important in determining the final chemical composition of the ejected gas than are the nuclear reactions which occurred during the star's evolution before the explosion. It seems possible but not certain that supernovae are sufficiently frequent to account for the rate at which heavy elements are being fed into the interstellar medium at the present time but there must have been a much higher rate of supernovae in the first generations of massive stars.

The production of helium in stars poses a problem if the Galaxy originally contained no helium, but there is no particular difficulty if most of the helium is primaeval. Although helium is produced from hydrogen in most stars, it is subsequently converted into heavier elements. It is very difficult to see how the observed ratio of helium to heavy elements can be produced, unless most stellar explosions leave behind rather large remnants, and this would imply that a large fraction of the mass of the Galaxy is now in the form of black holes. It is just possible that very massive stars could lose mass after

144

hydrogen has been converted into helium and before the heavy elements are produced and so avoid the difficulty concerning the relative production of helium and heavy elements.

Although the production of the heavy elements in stars seems quite plausible, the observations of explosions in the central regions of some galaxies, which obviously involve a much greater mass than that of a supernova, have led to suggestions that objects much more massive than ordinary stars might have played an important role in nucleosynthesis. There is, as yet, no very adequate theory of the explosions that do occur in galactic centres, although some workers have suggested that they might be bursts of creation of new matter rather than the explosion of a superstar. Calculations of nucleosynthesis in these little bangs have shown that both helium and heavy elements can be produced, but it is not at present believed that they are a major source of heavy elements because they appear not to produce the correct mix of heavy elements.

We have not discussed in any detail how the observed mix of heavy elements may have been produced, as this could only be done in a much longer book. We have, however, pointed out that the observation that the mix of heavy elements is very similar from star to star does not necessarily imply that one type of object has been responsible for all nucleosynthesis. The material ejected by any one supernova is so considerably diluted before it is part of the ordinary interstellar medium that a very large number of supernovae must influence a region of gas for its composition to change significantly and a uniform mix can be obtained by the whole of the interstellar gas being influenced by a representative sample of supernovae.

CHAPTER 8
radioactive chronologies

Introduction

IN this chapter we discuss the special role played by the natural radioactive elements in giving us information about the life history of the Universe and of the origin of the elements. We shall see that study of those elements with naturally occurring radioactive isotopes gives very reliable information about the ages of objects in the solar system and that it can also give useful, but less certain, information about the total timescale of element production and of time variations in this production.

Ages of Earth, Moon and meteorites

The best known application of the laws of decay of natural radioactive isotopes is the determination of the ages of terrestrial and lunar rocks and meteorites. It can be described by reference to isotopes of uranium and thorium, although some similar information can be obtained from other elements. The three isotopes ^{232}Th, ^{235}U and ^{238}U all decay to different isotopes of lead, ^{208}Pb, ^{207}Pb and ^{206}Pb. All of these isotopes have long half lives whose values are:

$$
\left.
\begin{array}{ll}
^{232}\text{Th} & 13{\cdot}9 \times 10^9 \text{ yr,} \\
^{235}\text{U} & 0{\cdot}7 \times 10^9 \text{ yr,} \\
^{238}\text{U} & 4{\cdot}5 \times 10^9 \text{ yr.}
\end{array}
\right\}
\qquad (8.1)
$$

The complete decay chain for ^{238}U is shown in Table 15. It can be seen that the decay consists of a mixture of α particle emissions, when the mass number decreases by four and the charge number by two,

$$
\begin{array}{l}
{}_{92}^{238}\text{U} \xrightarrow[\alpha]{4{\cdot}5 \times 10^9 \text{ yr}} {}_{90}^{234}\text{Th} \xrightarrow[\beta]{24{\cdot}1\text{d}} {}_{91}^{234}\text{Pa} \xrightarrow[\beta]{6{\cdot}7 \text{ hr}} {}_{92}^{234}\text{U} \xrightarrow[\alpha]{2{\cdot}5 \times 10^5 \text{ yr}} {}_{90}^{230}\text{Th} \xrightarrow[\alpha]{80 \text{ yr}} {}_{88}^{226}\text{Ra} \\[2em]
\xrightarrow[\alpha]{1{\cdot}6 \times 10^3 \text{ yr}} {}_{86}^{222}\text{Rn} \xrightarrow[\alpha]{3{\cdot}8\text{d}} {}_{84}^{218}\text{Po} \xrightarrow[\alpha]{3{\cdot}1\text{m}} {}_{82}^{214}\text{Pb} \xrightarrow[\beta]{26{\cdot}8\text{m}} {}_{83}^{214}\text{Bi} \xrightarrow[\beta]{19{\cdot}7\text{m}} {}_{84}^{214}\text{Po} \\[2em]
\xrightarrow[\alpha]{164\ \mu\text{s}} {}_{82}^{210}\text{Pb} \xrightarrow[\beta]{21 \text{ yr}} {}_{83}^{210}\text{Bi} \xrightarrow[\beta]{5{\cdot}0\text{d}} {}_{84}^{210}\text{Po} \xrightarrow[\alpha]{138\text{d}} {}_{82}^{206}\text{Pb}
\end{array}
$$

Table 15. The radioactive decay chain of ^{238}U. The half-life of each unstable isotope is shown together with the decay mechanism (α-decay or β-decay).

146

and β-decay, with a unit increase in charge number. In this decay chain all of the intermediate half lives are very much smaller than the half life of ^{238}U; the decay chains of ^{232}Th and ^{235}U have a similar property. This means that at any time in the decay of ^{238}U almost all of the material will be either ^{238}U or ^{206}Pb, the abundances of intermediate isotopes being negligible. The amount of lead present will gradually increase with time.

When we study a rock sample from either the Earth or the Moon or a meteorite we can hope to discover how long it is since the rock solidified. We can do this by studying the relative concentrations of uranium and thorium and of the isotopes of lead. When a rock solidifies, further *chemical separation* of its components becomes impossible, apart from a possible slight seepage of gases. This means that the uranium and thorium and any lead that has been produced by their decay will be closely associated.

Suppose that, in a unit mass of rock at the time of its solidification, there are $N_0(A)$ atoms of a radioactive isotope A which has a *decay rate**λ. Then the number of atoms of the isotope A changes with time according to the equation

$$\frac{dN(A)}{dt} = -\lambda N(A). \tag{8.2}$$

This equation can be integrated immediately to give

$$N(A) = N_0(A) \exp(-\lambda t), \tag{8.3}$$

where we are taking the zero of time at the time of solidification of the rock. If the decay product, isotope B, started with initial abundance $N_0(B)$, its abundance at time t would (clearly) be

$$N(B) = N_0(B) + N_0(A)(1 - \exp(-\lambda t))$$
$$= N_0(B) + N(A)(\exp \lambda t - 1). \tag{8.4}$$

where isotope B is assumed to be stable. Although there are intermediate isotopes in the decay chains of uranium and thorium to lead, their very short half lives mean that we can use formulae (8.3) and (8.4) with uranium or thorium as A and lead as B and with the decay rates corresponding to the half lives (8.1).

When we measure the present abundance of uranium, thorium and lead in a rock sample, we obtain the present abundances of ^{232}Th, ^{235}U, ^{238}U, ^{206}Pb, ^{207}Pb and ^{208}Pb and we have only three equations of the form (8.4) from which we can hope to find their initial abundances and the time which has elapsed since the rock solidified. How can we hope to make progress? The clue lies in the fact that *chemical separation* affects all isotopes of a given element similarly. If we assume that the solar system material was well mixed before

---

* Otherwise known as decay constant.

147

the Sun and planets were formed, the relative amounts of the different isotopes of uranium and lead should have been the same throughout the material in addition to the elemental abundances being uniform. This assumption that the solar system material was *well mixed* is supported by the similar isotopic ratios in elements, which have not been affected by radioactive decay, found in terrestrial, lunar and meteoritic samples, which have been mentioned on page 26. When the rocks and meteorites were formed the relative amounts of the elements uranium and lead may have been affected by chemical processes but all isotopes of any one element should have been affected similarly.

If we write the decay rates of ^{232}Th, ^{235}U and ^{238}U as λ_{232}, λ_{235} and λ_{238} and assume as mentioned above that the abundances of intermediate isotopes are small, we can write, using eqn. (8.4),

$$N(^{208}\text{Pb}) = N_0(^{208}\text{Pb}) + N(^{232}\text{Th}) \,(\exp \lambda_{232}t - 1), \qquad (8.5)$$

$$N(^{207}\text{Pb}) = N_0(^{207}\text{Pb}) + N(^{235}\text{U}) \,\,(\exp \lambda_{235}t - 1), \qquad (8.6)$$

and $\qquad N(^{206}\text{Pb}) = N_0(^{206}\text{Pb}) + N(^{238}\text{U}) \,\,(\exp \lambda_{238}t - 1). \qquad (8.7)$

In what follows we concentrate attention on the isotopes of uranium. Equations (8.6) and (8.7) can be combined to give

$$\frac{N(^{207}\text{Pb}) - N_0(^{207}\text{Pb})}{N(^{206}\text{Pb}) - N_0(^{206}\text{Pb})} = \frac{N(^{235}\text{U})}{N(^{238}\text{U})} \frac{(\exp \lambda_{235}t - 1)}{(\exp \lambda_{238}t - 1)} = f(t). \qquad (8.8)$$

In the expression which we have called $f(t)$, $N(^{235}\text{U})/N(^{238}\text{U})$ is the isotopic ratio of uranium *at the present time*, which is found to be 0·0072; this explains why we can write $f(t)$ as a function of t alone.

It is now convenient to express all of the lead abundances in terms of the abundance of an isotope, ^{204}Pb, which is not a radioactive decay product. Equation (8.8) can first be written

$$N(^{207}\text{Pb}) - N_0(^{207}\text{Pb}) = f(t)(N(^{206}\text{Pb}) - N_0(^{206}\text{Pb})). \qquad (8.9)$$

Division by $N(^{204}\text{Pb})$ then gives

$$\frac{N(^{207}\text{Pb})}{N(^{204}\text{Pb})} - \frac{N_0(^{207}\text{Pb})}{N(^{204}\text{Pb})} = f(t) \left[\frac{N(^{206}\text{Pb})}{N(^{204}\text{Pb})} - \frac{N_0(^{206}\text{Pb})}{N(^{204}\text{Pb})} \right], \qquad (8.10)$$

where $N(^{204}\text{Pb})$ could equally have been written $N_0(^{204}\text{Pb})$ as this abundance does not change with time.

Ages of meteorites

If it is now assumed that all of the meteorites were formed out of material with the same *isotopic* composition of lead and that the only thing which caused different meteorites to have different abundance ratios today was chemical separation of U and Pb at the time of formation, $N_0(^{207}\mathrm{Pb})/N(^{204}\mathrm{Pb})$ and $N_0(^{206}\mathrm{Pb})/N(^{204}\mathrm{Pb})$, which we cannot observe, should have been the same for all meteorites. The relation between $N(^{207}\mathrm{Pb})/N(^{204}\mathrm{Pb})$ and $N(^{206}\mathrm{Pb})/N(^{204}\mathrm{Pb})$ for a variety of meteorites should then be a straight line of the form

$$y - y_0 = m(x - x_0), \qquad (8.11)$$

whose slope m is $f(t)$, which determines the time of solidification of the meteorites and whose intercepts on the axes give the initial lead abundance ratios.

This procedure will make sense if all of the meteorites solidified at about the same time. The method has been applied to several meteorites of different types and the results do lies on a straight line indicating an age of about $4{\cdot}55 \times 10^9$ years*. The results are shown in fig. 56. Also shown are results corresponding to the oldest known terrestrial samples which give a similar age to that found for the meteorites. The accuracy of these results is really quite remarkable, the age being known to about 50 million years. It appears that something very significant happened about $4{\cdot}5 \times 10^9$ years ago.

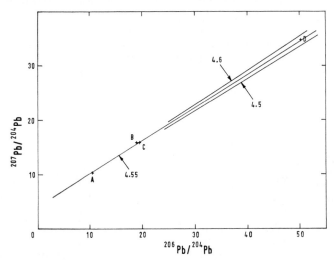

Fig. 56. Meteoritic ages. The lead isotopic abundances in A iron meteorites, B oceanic sediments, C chondritic meteorites, and D an achondrite lie on a line which indicates a common age of $4{\cdot}55 \times 10^9$ yr. The lines for $4{\cdot}5$ and $4{\cdot}6 \times 10^9$ yr are also shown.

* Very recently there have been discovered a very few meteorites which give a different and very much smaller age. This suggests that these meteorites must have been formed by the fragmentation of a more massive parent body in the recent (astronomical) past. This fragmentation must have been sufficiently violent for melting and chemical separation of the radioactive elements and lead to have occurred.

Various objects, including the parent bodies of the meteorites solidified then and it is generally believed that this marks the time of formation of the planetary system.

Ages of terrestrial and lunar rocks

We do not expect to obtain a single age for all of the Earth's rocks. There has been and is continuing volcanic activity on the Earth and the Earth's rocks have solidified at widely different times in geological history. There is ample opportunity for chemical separation up to the final solidification, so that we can expect to find widely different ages by study of the radioactive elements in the way we have described. This is found and most of the rocks are very much younger than the meteorites, most of the Earth's surface having been affected by relatively recent mountain building. The most significant result is that no terrestrial samples, which have been studied, have given greater ages than the meteorites. Only a very small fraction of the Moon's surface has been studied so far. The ages of the first rocks which were collected were found to be about $3 \cdot 5 \times 10^9$ years. We know almost nothing of the Moon's past geological (or more correctly selenological) history, but it is of great interest that the first rocks studied should be so very old. Later lunar samples have given even greater ages approaching those of the meteorites. The lunar surface has certainly had a much less disturbed life history than the terrestrial surface.

Use of helium to date rocks

It is perhaps of interest to describe the first method used to determine the age of a terrestrial rock by radioactive techniques. It is perhaps the most obvious method but unfortunately it does not give very reliable results. Lord Rutherford estimated the age of a piece of pitchblende to be 700 million years at a time when geologists believed that the Earth was no more than 100 million years old. He measured the amount of helium (α particles) trapped in the rock and assumed that this had all been produced by radioactive decay. This would give a very reliable age for the rock, if it could be assumed that all the helium produced would be trapped. As helium is gaseous, there is a distinct possibility that some of it can escape through pores in the rock and the method therefore gives an underestimate of the amount of radioactive decay that has occurred and hence of the rock's age.

All of the studies of solar system rocks are consistent with the idea that all objects in the solar system had a common origin and that the age of the solar system is about $4 \cdot 55 \times 10^9$ years. They do not, of course, prove that this is so but studies of the structure and evolution of the Sun also suggest that its present properties are consistent with its being of about the same age.

150

The age of the elements

We now turn to deductions from the abundances of the radioactive elements which are much less accurate. As mentioned earlier it is possible to use observations of the abundances of the radioactive elements and their decay products to obtain information about the *history* of element formation, at least in our neighbourhood in the Galaxy. To do this we must use theoretical ideas about the origin of the heavy radioactive elements which have been discussed in Chapter 5. It is believed that all of these isotopes have been produced by the *r*-process of neutron capture; even if we do not know where, in nature, the *r*-process occurs, there seems good reason for believing that it is the *basic* process. The theory of the *r*-process makes predictions of the relative production rates of the different isotopes. According to the results described in Chapter 5 and shown in fig. 35, the abundances of different isotopes produced by the *r*-process are fairly similar except near the magic number nuclei. Detailed calculations of the production of ^{232}Th, ^{235}U and ^{238}U in the *r*-process predict that amounts of the three isotopes produced should be in the ratios

$$\left. \begin{array}{l} N_0(^{235}\text{U}) \ / N_0(^{238}\text{U}) \approx 1\cdot 45, \\ N_0(^{232}\text{Th})/N_0(^{238}\text{U}) \approx 1\cdot 7, \end{array} \right\} \tag{8.12}$$

Assuming that the relative production of the isotopes is correct, it is possible to attempt to relate their present abundances to their production.

Knowing the present values of the ratios $N(^{235}\text{U})/N(^{238}\text{U})$ and $N(^{232}\text{Th})/N(^{238}\text{U})$, we can first ask, assuming that all of the material was produced in one event, how long is it since the synthesis occurred? We do this by assuming that, since the time of formation, the abundance of each isotope has decreased with its known decay rate. It is *a priori* unlikely that all of the radioactive elements in the solar system were formed simultaneously and the result of this calculation immediately gives an inconsistency. The $^{235}\text{U}/^{238}\text{U}$ ratio suggests that the synthesis must have occurred about $6\cdot 5 \times 10^9$ years ago but the $^{232}\text{Th}/^{238}\text{U}$ ratio indicates synthesis about 8×10^9 years ago. This really is a discrepancy greater than that allowed by uncertainties in the relative production rates, as the half life of ^{235}U, $0\cdot 7 \times 10^9$ yr, is small compared to the difference in the two times.

This calculation can be supplemented by assuming that synthesis in the solar system material lasted for an interval of time T before the formation of the Earth and the meteorites. During that time we assume that the rate of production of the heavy elements declined according to the law $\exp(-\Lambda t)$, where Λ is initially arbitrary and calculations can be made for a variety of values of Λ. The reason for assuming a declining rate of element production is that we have the evidence described in Chapter 3 that only the oldest stars are very short of heavy elements and that there must initially have been a very rapid production of them; the exponential law is only meant to be a first approximation to the truth. In the exponential law, $\Lambda = 0$ corresponds to a uniform rate of synthesis and $\Lambda \to \infty$ to sudden synthesis in a single event.

It is possible to ask whether any choice of Λ and T will give consistent results for the abundances of all three isotopes at the time of formation of the solar system. The results of calculations are shown in fig. 57, and it is found that, if there was a gradually declining rate of heavy element production in the Galaxy over a period of about $6 \cdot 5 \times 10^9$ years *before* the solar system was formed, the abundances are approximately concordant. The accuracy of this figure must not be exaggerated, but the estimated time of commencement of heavy element synthesis (11×10^9 years ago) is not inconsistent with other estimates of the age of the *Galaxy*.

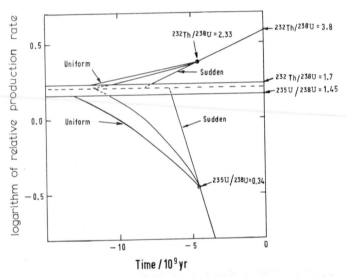

Fig. 57. The time of formation of heavy radioactive elements. The present abundances of U and Th can be used to deduce the abundance ratios $^{232}\text{Th}/^{238}\text{U} = 2 \cdot 33$ and $^{235}\text{U}/^{238}\text{U} = 0 \cdot 34$ when the solar system was formed. If the heavy elements were produced in a single event, abundances earlier than that follow the curves marked 'Sudden'. If there was a uniform rate of nucleosynthesis the curves marked 'Uniform' apply. Neither of these assumptions lead to the two ratios having their *r*-process production values at the same time. The unmarked curves corresponding to a declining rate of nucleosynthesis do yield a consistent age of 11×10^9 yr. As the *r*-process production rates are uncertain, the dashed lines give corresponding results if the initial values of $^{232}\text{Th}/^{238}\text{U}$ and $^{235}\text{U}/^{238}\text{U}$ are taken to be equal at $1 \cdot 6$.

The past rate of nucleosynthesis

It is possible to obtain additional information about the timescale of formation of the radioactive elements if a few plausible assumptions are made. If nucleosynthesis has occurred continuously, but not necessarily at a uniform rate in a system of constant mass (the Galaxy), it is possible to write down equations governing the abundances of nuclei as a function of time. Suppose we consider a radioactive isotope which is produced by the r-process and is then ejected into the interstellar medium and allowed to decay. For isotopes of very long half life, such as ^{232}Th, very little decay can have occurred and most of what has been produced must still be in the same form. Nuclei of very much shorter half life will have largely decayed. Consider the rate of change with time of the abundance of such a radioactive isotope. This can be written

$$\frac{dN_i}{dt} = P_i - \lambda_i N_i, \tag{8.13}$$

where N_i is the abundance of the isotope, P_i its rate of production and λ_i its decay rate.

If we consider an isotope of very short half life, it will rapidly approach a situation where its rate of production is balanced by its rate of decay. Provided that the production rate is slowly varying compared to the timescale λ_i^{-1}, we obtain

$$N_i(t) = P_i(t)/\lambda_i. \tag{8.14}$$

In contrast, a stable or very long lived isotope will have the abundance

$$N_j(t) = \int_0^t P_j(t')dt' = t\langle P_j \rangle, \tag{8.15}$$

where $\langle P_j \rangle$ is the mean rate of production of the stable isotope. If it is assumed that both isotopes have been produced by the r-process so that, as described in Chapter 5, their relative rates of production can be predicted by theory, we have a value for $P_i(t)/P_j(t)$. The ratio of the two abundances at time t can be written

$$\frac{N_i(t)}{N_j(t)} = \frac{P_i(t)}{P_j(t)} \frac{P_j(t)}{\lambda_i t \langle P_j \rangle}. \tag{8.16}$$

As λ_i and $P_i(t)/P_j(t)$ are known, eqn. (8.16) expresses the ratio of the two abundances in terms of the time for which nucleosynthesis has been proceeding, t, and the present rate of nucleosynthesis divided by its mean rate in the past. If we could observe the ratio of abundances in the interstellar medium today, we could insert the estimate of the total timescale of nucleosynthesis obtained from the study of uranium and thorium described above into eqn. (8.16) and obtain a value for the ratio of the present rate of nucleosynthesis to its mean rate in the past.

This argument does not, however, describe the observations which we make. We are not able to study the abundances of short-lived radioactive isotopes and stable isotopes in the interstellar gas *today*. Instead we observe the stable isotopes and the decay products of the radioactive isotopes in the meteorites. The meteoritic material has not been affected by nucleosynthesis during the past 4.55×10^9 yr, since the meteorites solidified. Let us suppose that this solidification occurred at a time T after the start of nucleosynthesis; we have previously estimated T to be 6.5×10^9 yr. From the observations of the decay products of radioactive isotopes, we can estimate the abundance of the radioactive isotope when the meteorite solidified and we know the abundance of the stable isotope. By applying eqn. (8.16) at time T, we then obtain a value for $P_j(T)/T\langle P_j \rangle$.

Nucleosynthesis in solar system material

This calculation can be performed for several short-lived isotopes but it gives inconsistent results. The results can be made consistent if it is assumed that there was a period of about 10^8 yr before the meteorites solidified during which the solar system material underwent no further nucleosynthesis and the radioactive isotopes decayed freely; the short-lived isotopes would decay substantially in such a time. If this result is confirmed by further investigations, it will be necessary to try to decide whether there was genuinely no nucleosynthesis in the part of the Galaxy containing the Sun in a short period before the solar system was formed or whether some process shielded the solar system material from further enrichment with heavy nuclei.

If the solution of eqn. (8.13) is studied for radioactive isotopes whose half life is small compared to the age of the Galaxy but longer than that of the isotopes just discussed, further information can be obtained about the past rate of formation of the heavy elements. The detailed discussion is far too complicated to reproduce here but one result which is obtained from study of both these and the short-lived isotopes can be reported. If we take the time T to be the time at which nucleosynthesis ceased in the material out of which the solar system was formed, it appears that

$$(P_j(T)/\langle P_j \rangle) > 1. \qquad (8.17)$$

This means that the rate of nucleosynthesis just before the solar system was formed was higher than the average rate throughout the previous life history of the Galaxy. The formation of the solar system was apparently preceded by a high rate of nucleosynthesis and it is possible that the explosion of a number of supernovae and the interaction of their ejected gas with the existing interstellar medium sets up conditions which are particularly favourable for the onset of star formation. This must certainly be investigated further.

To conclude this discussion of radioactive chronologies, one point should be stressed. All of our observations refer to *solar system material*. It is only possible to make direct deductions about the progress of nucleosynthesis in this solar system material. When we refer to the age of the radioactive elements and when we compare the rate of formation of the heavy elements just before the solar system was formed with the average rate of formation throughout galactic history, we are only making statements about the solar system material. We may wish to generalize the results which have been obtained to the Galaxy at large but, if we do that, we must realise that we are making some assumptions about the uniformity on a large scale of conditions in the Galaxy. Such assumptions would presumably be justified by remarking that the relative amounts of the heavy elements in different young stars and interstellar gas clouds appear to vary by no more than a factor of about two in that part of the Galaxy, which can be studied in detail. This may indicate that the past life history has been generally the same throughout at least a large part of the Galaxy. Ultimately the variations in the past rate of nucleosynthesis deduced from study of the radioactive isotopes must be compared with the knowledge of the rate of nucleosynthesis gained from studies of type described on pages 136 and 137 of Chapter 7.

Summary of Chapter 8

Useful information about the ages of objects in the solar system can be obtained by a study of the naturally occurring radioactive isotopes. After a rock solidifies, radioactive elements such as uranium and thorium and the lead to which they decay remain associated together. Since chemical separation affects all isotopes of a given element similarly, it may be assumed that the isotopic composition of lead was the same in all rocks which solidified at the same time. A study of the present isotopic composition of lead and the relative abundances of uranium and thorium to lead enables a rock's age to be found. The meteorites have an age of 4.55×10^9 yr. Terrestrial and lunar rocks tend to have lower ages than the meteorites, although the oldest samples are of comparable age. Typically, lunar rocks are much older than terrestrial rocks, indicating that the Earth has had a much more active geological history. It appears that the age of the solar system is about 4.55×10^9 yr.

By combining the observations of radioactive isotopes and their decay products with the assumption that they were produced by the r-process of neutron capture, so that relative production rates of different isotopes are known, it is possible to obtain some information about time variations of the rate of nucleosynthesis. *The results obtained are much less reliable than those concerned with the age of the*

CHAPTER 9
concluding remarks

IN this book we have shown how observations of the chemical composition of objects in the Universe can be combined with theoretical ideas about how the different elements can be produced to explain how the chemical composition of the Universe has changed during its life history. Although we believe that the main outlines of the investigation are well-defined, there are uncertainties at each stage of the discussion. For this reason our main purpose has not been to present detailed results which are incomplete but to indicate the way in which the problem is being approached. Even the achievement of this limited aim is not particularly easy. Different branches of physics and astronomy interact in a very complicated way, and, if this book does not possess a clear logical development, it is because this is an intrinsic property of the subject. The inter-relations between cosmology, stellar and galactic evolution, geochemistry, atomic and nuclear physics will only become completely clear when a reasonably complete solution of the problem has been obtained. In this chapter, we mention some aspects of our treatment of the subject which will need further investigation before this complete solution can be obtained.

The main question discussed in this book is whether the heavy elements, and possibly also the helium, that are observed in the Galaxy today can have been *produced in stars* during the galactic lifetime. It is clear that estimates of the probability that this is true depend on:

    (i) the age of the Galaxy;

    (ii) the rate at which stars evolve;

    (iii) the nucleosynthesis occurring inside stars;

    (iv) the mass loss from stars;

    (v) the numbers of stars of different types.

In our discussion we have concentrated attention on points (iii), (iv) and (v) and have assumed that (i) and (ii) are relatively more certain. How true is this?

The age of the Galaxy, which is believed to be something in excess of 10^{10} years, is estimated in two different ways. One depends on the ages of the globular clusters, which can be obtained by using (ii), and on ideas about the formation and flattening of the Galaxy. These suggest that the globular clusters must be almost as old as the Galaxy itself. The second method involves an estimate of the age of the Universe based on the present expansion rate and a cosmological theory, and the Galaxy is then required not to be older than the Universe; this method cannot be used in conjunction with a cosmological theory which says that the Universe has infinite age. The estimates of galactic age obtained by a study of globular clusters and from the big-bang cosmological theory are sufficiently close that it is generally believed that the age of the Galaxy is known to within an uncertainty of order 50 per cent. The rate at which stars of a given mass and chemical composition evolve, which has been discussed in *The Stars*, is thought to be known to a greater accuracy than the galactic age.

The laws of physics

Thus it *appears* that we are justified in regarding (i) and (ii) as relatively well known in comparison with the uncertainties which arise in (iii), (iv) and (v), but there is just one important reservation. Estimates of the age of the Galaxy and the rate at which stars evolve depend on the assumption that the laws of physics which we determine today have been applicable at *all points* in the Universe *throughout its life history*. We have already mentioned in Chapter 6 that this assumption may not be correct and that theories have been postulated in which quantities such as the ratio of the gravitational to the electrostatic forces between elementary particles have not always had the same value. As the luminosity of a star of given mass depends on the gravitational constant to a very high power ($L_s \propto G^7$ for a star of mass $\approx 1\ M_\odot$), estimates of stellar and galactic ages could be very wrong, if such a theory were correct. Consideration such as this *could* have an important effect on the discussion in this book but, having given a warning that the laws of physics *may* not be unvarying, we will now turn to some outstanding problems within the framework of the existing laws of physics.

The helium problem

It should be clear from Chapters 6 and 7 that possibly the greatest difficulty in a study of the origin of the elements is the uncertainty about the helium content of the Universe. It is possible that all objects in the Universe have a helium content of order 25 per cent by mass and that this helium could have been present before any galaxies were formed, but on the basis of the present observations it is also possible that the galaxies contained no helium at their time of formation. If there were no other important uncertainties in the theories and observations, it would be feasible to demonstrate that one or other of these initial conditions is impossible, but at present that cannot be done. Clearly any further information that can be obtained about the helium content of objects in the Universe will be very valuable in helping to remove this great uncertainty.

The origin of the first metals

Throughout this book we have assumed that the initial chemical composition of the Universe was either pure hydrogen or a mixture of hydrogen and helium; by implication we have also assumed that the Galaxy contained no metals when it was formed. Although the oldest known stars, which are members of globular clusters, do have an extremely low metal content, they contain a measurable quantity of metals. We must therefore ask why the oldest stars which we observe contain metals or alternatively why do we not observe stars with no metals? If the Galaxy really contained no heavy elements initially, it seems necessary to suppose that there was a first generation of massive metal-free stars which evolved rapidly before the globular clusters were formed and which provided the small quantity of metals observed in the globular cluster stars; it is necessary that there should not be too many low mass stars in this generation, because they would not yet have evolved and would be around today and possess no metals. It is interesting that this argument for a first generation of massive stars is additional to those already given in Chapter 7.

Recent observations of globular clusters have indicated that not all clusters have the same metal content; those clusters that move high above the galactic plane contain less metals than those which always remain close to the plane. This suggests that the earliest metals may have been formed preferentially near the plane and the centre of the Galaxy and also that the globular clusters may not have been formed in the initial collapse of the Galaxy but instead during an outward bounce after the first collapse. These questions all require further study.

The intergalactic medium

In our discussion of the chemical composition of the Galaxy, it has been regarded as an isolated system. We have mentioned in passing that there might be *intergalactic* matter and that the cosmic rays eventually escape into intergalactic space. We have also said that, if there is insufficient interstellar matter, the mass ejected by galactic explosions can escape from the Galaxy, but we have not mentioned the alternative possibility that the Galaxy could accrete a significant amount of intergalactic matter. Some observations suggest that there may be an important exchange of mass between the Galaxy and the intergalactic medium. High velocity clouds of gas appear to be falling towards the plane of the Galaxy, but it is not entirely clear whether they contain genuine intergalactic matter or whether they contain material that has previously been thrown high above the galactic plane by explosions in the Galaxy. The possible influence of mass exchange between the Galaxy and the intergalactic medium on the chemical composition of the interstellar gas needs to be added to the effects already discussed in this book.

Models of galactic nucleosynthesis

In Chapter 7 on page 137, we have described how, by making assumptions about the rate of star formation and the extent of nucleosynthesis and mass loss in stars of different masses, it is possible to calculate the chemical composition of the Galaxy as a function of time and to compare the results with observation. As calculations of stellar evolution and mass loss are refined, it should be possible to improve the initial assumptions of these models. The calculations which have been done at present treat the Galaxy as a uniform medium. Now that observations of spatial variations of chemical composition in the Galaxy are being accumulated, it will soon be desirable to make calculations of the rate of galactic nucleosynthesis which include this spatial structure. This will be considerably more difficult than the present calculations, because not only must the rate of nucleosynthesis be allowed to vary from place to place in the Galaxy but the diffusion of interstellar gas from one part of the Galaxy to another must also be included.

Limitations of the present discussion

In conclusion it should be stressed again that we have only been able to discuss in broad outline the vast amount of work that has been done on this subject. Because a somewhat superficial treatment has been forced upon us by limitations of space, we may have given the impression that the subject has so far only been treated superficially. *This is certainly not true.*

The analysis of the spectrum of only one star to produce a reliable chemical composition is a very lengthy task but many abundance determinations are now being published each year. The calculation of rates of nucleosynthesis in an evolving, and possibly exploding, star is a difficult computational problem, but the mathematical models being used are becoming more realistic every year. The major recent development on the observational side has been the demonstration that place of origin may be of comparable importance to age in determining the chemical composition of a star. A significant theoretical development has been the realization that explosive nuclear reactions may have much more influence on the chemical composition of the interstellar gas than fusion reactions which occur in the quiet stages of stellar evolution. Although we have concentrated our discussion on the total production of helium and heavy elements in

160

the life history of the Galaxy, when theoretical and observational astronomers meet to discuss their results, they are much more likely to be asking whether or not a process produces a *particular* group of isotopes than to be discussing the total rate of nucleosynthesis. *Ultimately, the success of any theory of the origin of the chemical elements will depend on its ability to give a full explanation of the abundances of all of the elements.*

Such a discussion of the abundances of individual elements and isotopes would not have been possible in this book without a considerable increase in both its length and complexity. The purpose of writing this book will, however, have been met if we have succeeded in conveying, in general terms, the fascination of the subject and if a small number of our readers are stimulated to a study of the more detailed problems which we have omitted. Although the broad outlines of the subject are clear, there will be much detailed work to be done for many years yet.

APPENDIX

thermodynamic equilibrium*

IF a physical system is isolated and left alone for a sufficiently long time, it settles down into what is known as a state of thermodynamic equilibrium. In thermodynamic equilibrium the overall properties of the system do not vary from point to point and do not change with time. Individual particles of the system are in motion and do have changing properties. For example, electrons may be being removed from and attached to atoms. There is, however, a statistically steady state in which any process and its inverse occur equally frequently. Thus, in the example mentioned above, the number of atoms ionized per unit time is equal to the number of recombinations. Because the properties of a system do not vary from point to point when it has reached thermodynamic equilibrium, all parts of it have the same temperature.

If two such isolated systems are brought into contact, heat will flow from one to the other until they reach the same state of thermodynamic equilibrium and hence the same temperature. In thermodynamic equilibrium all of the physical properties of the system (such as pressure, internal energy, specific heat) can be calculated in terms of its density, temperature and chemical composition alone. In nature a true state of thermodynamic equilibrium may be approached closely but never quite reached.

In thermodynamic equilibrium the intensity of radiation is given by the Planck function

$$I_\nu = B_\nu(T) \equiv \frac{2h\nu^3/c^2}{\exp(h\nu/kT) - 1}. \qquad (A.1)$$

It should be noted that the Planck function is determined by the temperature alone and does not depend on the density and chemical composition of the material. Provided corrections due to quantum mechanics and relativity are slight (and this is often true), any species of particle possesses a Maxwellian velocity distribution

$$f(u, v, w) = n\left(\frac{m}{2\pi kT}\right)^{3/2} \exp\left(-m(u^2 + v^2 + w^2)/2kT\right), \qquad (A.2)$$

where n is the number of the particles in unit volume, m the particle mass, u,v,w the three components of velocity of the particles and the

* This is a modified and extended version of an Appendix which appeared in *The Stars*.

162

number of particles in unit volume with velocities between (u,v,w) and $(u + \delta u,\ v + \delta v,\ w + \delta w)$ is $f\ \delta u\ \delta v\ \delta w$.

In any particular atom or ion, electrons will be arranged in various energy levels. If two such levels have energies E_r and E_s, in thermodynamic equilibrium the numbers n_r, n_s of atoms in the two states will obey the Boltzmann law

$$\frac{n_r}{n_s} = \frac{\exp\left(-E_r/kT\right)}{\exp\left(-E_s/kT\right)}. \tag{A.3}$$

Finally an atom can exist in several states of ionization and the number of atoms per unit volume in two successive states of ionization, $n_i{}^*$, n_{i+1} is related to the electron density n_e by Saha's equation

$$\frac{n_{i+1}n_e}{n_i} = \left(\frac{2\pi mkT}{h^2}\right)^{3/2} \frac{2B_{i+1}}{B_i} \exp\left(-I_i/kT\right), \tag{A.4}$$

where I_i is the energy required to remove one electron from the atom in the ith state of ionization and B_i, B_{i+1}, which are called the partition functions for the two states, depend on the electron energy levels in the two ions and the temperature. The chemical composition of the medium enters into Saha's equation because the electrons entering into n_e can be provided by ionization of all of the elements present.

In the deep interior of a star departures from thermodynamic equilibrium are slight but, as the surface of the star is approached, (A.1) at least must cease to be true. For one reason, the radiation primarily flows outward and in addition the distribution of radiation with frequency ceases to have the Planck form. If collisions between particles are sufficiently rapid, (A.2)–(A.4) may still continue to be valid for some quantity T, which is not a true thermodynamic temperature but which is known as the kinetic temperature. If this is true the system is said to be in a state of *local thermodynamic equilibrium*. When population of the states of excitation and ionization follow (A.3) and (A.4), interpretation of properties of spectral lines in terms of element abundances is relatively straightforward. Very near the surfaces of stars it seems certain that the approximations of local thermodynamic equilibrium break down and it is then much more difficult to convert observed line strengths into abundances. This is particularly true of regions where emission lines are formed and that is why emission lines rarely yield reliable abundances. In the interstellar medium conditions are very far from thermodynamic equilibrium.

In the above discussion it has been assumed that the chemical composition of the material can be specified but, in fact, that is not strictly true. Nuclear reactions gradually convert nuclei from one form to another. Usually nuclear reactions occur so slowly that the

* The quantities in (A.3) should now be written more correctly n_{ir}, n_{is}.

assumption of unchanging chemical composition is justified in discussing the approach to thermodynamic equilibrium. In conditions of sufficiently high temperature and density nuclear reactions may occur so rapidly that an approach to nuclear statistical equilibrium takes place. If this happens the chemical composition of the material can no longer be specified in advance but it must be determined by using a set of Saha-type equations, which couple the densities of the free protons and neutrons and the complex nuclei. In complete thermodynamic equilibrium the properties of a system are determined by its density and temperature alone. It is possible that a close approximation to complete thermodynamic equilibrium has occurred in some parts of the Universe during its lifetime.

INDEX

Accretion   43
Ages of Earth, Moon and meteorites
   146–150
Ages of the elements   151–152
Antimatter   4

Black hole   13, 115, 122, 130, 136
Big-bang theory   6, 99–103, 138

Chemical separation of elements   21, 22
Continuous creation   99
Cosmic rays   55–64
Cosmic rays, break up   61, 62
Cosmic rays, energy   55, 63, 134
Cosmic rays, life history   59–62
Cosmic rays, origin   56–58, 117, 134
Cosmic rays, solar   47
Cosmological models, see big-bang theory, steady state theory, oscillating cosmology
Crab nebula   57, 58, 115, 123
Cross-section   35, 74

Degenerate gas   120, 121
Doppler effect   30, 95, 115

Earth, atmospheric transparency   44
Earth, core   19–22
Earth, magnetic field   20, 21
Earth, mantle   19–21, 24, 25
Earth, structure of   19–21
Earthquake waves   20, 21
Effective temperature   10–12, 35
Element abundances, composite   13, 53
Element abundances, cosmic rays   56, 62
Element abundances, galaxies   52
Element abundances, interstellar medium   51, 52
Element abundances, solar,   44, 45
Element abundances, solar system   27, 78, 81, 89, 140, 141
Element abundances, young stars   52
Element production, cosmological   93, 101–103

Element production, massive objects   137–140
Element production, rate of   49, 128, 152
Element production, stars   109 *et seq.*
Elements, atmophile   22
Elements, chalcophile   22, 24
Elements, lithophile   22, 24
Elements, siderophile   22, 24
Evolved stars, chemical composition   113, 114
Expansion of the Universe   95, 98

f-value   35, 36, 38, 40, 42, 45

Galactic centres, explosions in   15, 16, 57–59, 138
Galactic chemical composition, spatial inhomogeneities   14, 16, 50, 108, 127, 138, 159
Galactic dimensions   9
Galactic disk   8, 9, 48, 127, 134
Galactic halo   9, 48, 127
Galactic magnetic field   60, 61
Galactic mass   9
Galactic nucleosynthesis, models of   136, 137, 160
Galaxies, formation of   105, 106
Galaxies, homogeneous distribution   96
Galaxies, isotropic distribution   95, 96
Galaxies, recession of   95, 100, 101
General relativity   100, 138
Giants   10, 124

Heavy element abundances   47–49, 108, 109
Heavy element production   110, 117, 132–136, 139, 140, 152
Heavy radioactive elements   18, 146, 151, 152
Heisenberg's principle of uncertainty   37
Helium abundance   44, 47, 49, 104, 105, 108, 113, 158

165

Helium production 102, 104, 129–132, 139
Hertzsprung-Russell diagram 10, 11
Hubble's constant 95

Intergalactic matter 101, 106, 159
Interstellar dust 10, 96, 132
Interstellar gas 10, 43, 96, 126, 127, 132
Interstellar gas, HI region 51
Interstellar gas, HII region 51
Interstellar gas, organic molecules 51
Iron-helium phase change 88, 89, 121, 143
Isotopes 3, 18, 26, 78
Isotopic abundances 26, 27, 88, 148

Jeans's instability 106
Jupiter, chemical composition 26

Lithium, beryllium and boron 46, 54, 62, 90, 91, 143
Little bangs 138–140
Luminosity 11, 12

Mach's principle 95
Main sequence stars 10, 111
Mass-luminosity relation 10, 12, 111, 112
Metals (see heavy elements)
Meteorites, carbonaceous chondrites 23
Meteorites, iron 23
Meteorites, stone 23–25
Meteorites, stony iron 23
Microwave background 98, 103–105
Moon, samples 24
Moon, seismic properties 24
Motion of charged particles 60

Neutrino reactions 88
Neutron capture abundances, r-process 81
Neutron capture abundances, s-process 77–79
Neutron capture cross-section 73–75
Neutron capture reactions 69–81
Neutron capture reactions, r-process 70–73, 79–81, 142, 143, 151
Neutron capture reactions, s-process 70–73, 75–79, 142, 143
Neutron exposure 77
Neutron star 12, 114, 117, 122, 123, 130
Nitrogen abundance 142
Novae 123

Nuclear binding energy 67, 68, 79
Nuclear fission reactions 67, 68
Nuclear fusion reactions 68, 69, 82–90, 111
Nuclei, bypassed 73, 78
Nuclei, magic number 68, 73, 77, 80, 151
Nucleogenesis (see element production, cosmological)
Nucleosynthesis (see element production, massive objects and stars)
Nucleosynthesis, past rate of 153, 154
Nucleosynthesis, solar system material 154, 155

Origin of the elements, $\alpha\beta\gamma$ theory 65, 66, 103
Origin of the elements, equilibrium theory 66
Origin of the elements, polyneutron theory 66, 67
Oscillating cosmology 99, 100, 102 103

Periodic table 1, 2
Physics, laws of 94, 158
Planck distribution 31, 51, 103
Planetary nebulae 124, 125
Planets, major 26
Planets, terrestrial 25, 26
Proton capture reactions 82
Pulsars 114, 122, 123

Quasars 97, 105

r-process (see neutron capture)
Radioactive chronologies 146–156
Radio galaxies 15, 58, 59, 97, 105
Radio galaxies, distribution 97
Radio source counts 97
Red shift 94, 95, 97, 98, 103

s-process (see neutron capture)
Snowplough mechanism 126
Solar nebula 22, 25
Solar wind 124
Spallation nuclear reactions 91, 143
Spectral classification 33–35
Spectral lines 29, et seq.
Spectral lines, absorption 29, 31, 32
Spectral lines, broadening 36–39
Spectral lines, curve of growth 39, 40, 42
Spectral lines, emission 29, 30, 45, 52

Spectral lines, equivalent width
   38–40, 42
Spectral lines, hydrogen 32, 33
Spectral lines, profile 36–39
Spiral arms 10, 50
Star clusters, galactic 10, 49
Star clusters, globular 8, 10, 48, 49,
   157, 159
Stars, high velocity 48
Stars, metal deficient 48
Statistics of element production
   127–137
Steady state theory 6, 99–101, 138
Stellar age 47
Stellar atmosphere, structure of 41
Stellar lifetimes 112
Stellar mass function 135, 136
Stellar mass loss 43, 114–125
Stellar place of origin 47, 49
Stellar winds 124
Supergiants 10
Supernovae 14, 15, 57, 115–123
Supernovae, frequency of 132, 133
Supernovae, invisible 119, 134
Supernovae, light curves, 115, 116
Supernovae, mass loss 117, 118, 133
Supernovae, remnants 115, 116, 122
Supernovae, theoretical models
   118–122
Supernovae, type I 116, 118
Supernovae, type II 116, 118
Synchrotron radiation 57, 117

Technetium 79
Thermodynamic equilibrium 29, 51,
   66, 162–164
Thermonuclear reactions 82–90, 111
Thermonuclear reactions, carbon
   burning 85, 121, 141, 142
Thermonuclear reactions, equilibrium
   86–89
Thermonuclear reactions, explosive
   84, 90, 110, 117, 119, 121, 122, 141,
   142
Thermonuclear reactions, helium
   burning 84, 85
Thermonuclear reactions, hydrogen
   burning 83, 84
Thermonuclear reactions, oxygen
   burning 85, 141, 142
Thermonuclear reactions, silicon
   burning 86, 87, 141, 142
Timescale, dynamical 71, 75, 118
Timescale, nuclear 71
Timescale, thermal 71, 118
Transuranium elements 56, 63, 80

Ultra-violet excess 42

Waiting point 80
White dwarfs 10, 114, 117, 122, 125
   130

X-ray background 98

Zone of avoidance 95, 96

SUGGESTIONS FOR FURTHER READING

THE subject is so broad that almost any book on stars, galaxies or cosmology will have some connection with it. It is not easy to find elementary books (or in some cases any book at all) on some aspects of the problem. In the following list the very advanced texts are marked with an asterisk.

For a general discussion the following books are recommended:

W. A. FOWLER, *Nuclear Astrophysics* (American Philosophical Society).
This is a non-mathematical account of the theoretical side of the subject and cosmochronology.

F. HOYLE, *Frontiers of Astronomy* (Heinemann, Mercury (paperback)).
This semi-popular book covers a much wider field than nuclear astrophysics. Many of its details are out of date, but it gives a lively picture of modern developments in astronomy.

*L. H. ALLER, *The Abundance of the Elements* (Interscience).
This advanced text is mainly concerned with the observations but includes a brief introduction to theoretical ideas.

*D. D. CLAYTON, *Principles of Stellar Evolution and Nucleosynthesis* (McGraw Hill).
This is another advanced text which discusses all of the basic nuclear processes.

R. J. TAYLER and A. S. EVEREST, *The Stars: Their Structure and Evolution* (Wykeham (paperback)).
This is a general introduction to the study of stellar evolution.

Books on specific topics are:

The Solar System

L. H. AHRENS, *Distribution of the Elements in our Planet* (McGraw Hill (paperback)).
This is a discussion of how the chemical composition of the Earth is determined.

B. MASON and W. G. MELSON, *The Lunar Rocks* (Wiley—Interscience).
This discusses the knowledge of the structure and chemical composition of the Moon obtained from the earliest Apollo missions.

Cosmology and the Universe

G. GAMOW, *The Creation of the Universe* (Mentor (paperback)).
An account of the $\alpha\beta\gamma$ theory of the origin of the elements and of the big-bang cosmological theory.

E. P. HUBBLE, *The Realm of the Nebulae* (Dover (paperback)).
An early account of the discovery of the distances to the galaxies and of their redshifts.

D. W. SCIAMA, *Modern Cosmology* (Cambridge)
A discussion of recent astronomical observations which are probably of cosmological significance.

D. W. SCIAMA, *The Physical Foundations of General Relativity* (Heinemann).
An introduction to general relativity in which the amount of mathematics used is kept to a minimum.

The following books in the Wykeham series are on related subjects:

G. E. BACON and G. R. NOAKES, *Neutron Physics*.

H. R. HULME and A. McB. COLLIEU, *Nuclear Fusion*.

G. M. LEWIS and G. A. WHEATLEY, *Neutrinos*.

Two key papers which played a major role in the recent development of the subject are:

*E. M. BURBIDGE, G. R. BURBIDGE, W. A. FOWLER and F. HOYLE, *Synthesis of the Elements in Stars*, Reviews of Modern Physics, **29**, 547, 1957.

*H. E. SUESS and H. C. UREY, *Abundances of the Elements*, Reviews of Modern Physics, **28**, 53, 1956.

Many additional references can be found in *Synthesis and Abundances of the Elements*, *Selected Reprints*, published for the American Association of Physics Teachers by the American Institute of Physics.

THE WYKEHAM SCIENCE SERIES

1 *Elementary Science of M*   I. W. MARTIN and R. A. HULL
2 *Neutron Physics*   G. R. NOAKES
3 *Essentials of Meteorolog*   S. THOM and
   (*Paper and Cloth Editi*   T. SAUNDERS
4 *Nuclear Fusion*   McB. COLLIEU
5 *Water Waves*   N. F. BARBER and G. GHEY
6 *Gra*   T. SAUNDERS
7 *Rel*
8 *The*   McCULLOCH
9 *Intr*   EASTWOOD
10 *The*   ARCHENHOLD
11 *Su*   S. EVEREST
12 *Ne*   R. NOAKES
13 *Cry*   WHEATLEY
14 *Bio*   R. M. LEE
15 *Un*   G. R. NOAKES
16 *Th*   R. TRICKER
17 *Th*   J. P. WEBB
18 *Ele*   T. RICHARDS
19 *Pr*   G. FOXCROFT
  (M. I. SMITH
20 *So*
21 *St*   J. SUTCLIFFE
22 *El*   R. A. HULL
23 *T*   NEVILL MOTT
                       S. EVEREST

THE ... IES

1 *F*   I. J. BEESLEY
2 *E*   HANDSCOMBE
3 *I*   HALMSHAW
4 *U*   H. WALLACE

All or... e appropriate
agents ... so of the title
page ...